Micropolitics and Canadian Business

Artist: Eugen Mihaescu

Micropolitics and Canadian Business

Paper, Steel, and the Airlines

PETER CLANCY

broadview press

Library and Archives Canada Cataloguing in Publication

Clancy, Peter, 1949–
    Micropolitics and Canadian business : paper, steel, and the airlines / Peter Clancy.

Includes bibliographical references and index.
ISBN 1-55111-570-0

    1. Wood-pulp industry—Canada.   2. Paper industry—Canada.
3. Steel industry—Canada.   4. Airlines—Canada.   I. Title.

HD2809.C48 2004   338.4'0971   C2004-903490-1

Broadview Press, Ltd. is an independent, international publishing house, incorporated in 1985. Broadview believes in shared ownership, both with its employees and with the general public; since the year 2000 Broadview shares have traded publicly on the Toronto Venture Exchange under the symbol BDP.

We welcome any comments and suggestions regarding any aspect of our publications—please feel free to contact us at the addresses below, or at broadview@broadviewpress.com / www.broadviewpress.com

North America
Post Office Box 1243,
Peterborough, Ontario,
Canada K9J 7H5
Tel: (705) 743-8990
Fax: (705) 743-8353
customerservice
    @broadviewpress.com

3576 California Road,
Orchard Park, New York
USA 14127

**UK, Ireland, and
Continental Europe**
NBN Plymbridge
Estover Road
Plymouth PL6 7PY
United Kingdom
Tel: +44 (0) 1752 202301
Fax: +44 (0) 1752 202331
Customer Service:
cservs@nbnplymbridge.com
orders@nbnplymbridge.com

**Australia and
New Zealand**
UNIREPS
University of
New South Wales
Sydney, NSW, 2052
Tel: + 61 2 96640999
Fax: + 61 2 96645420
info.press@unsw.edu.au

Broadview Press Ltd. gratefully acknowledges the financial support of the Government of Canada through the Book Publishing Industry Development Program for our publishing activities.

Cover design and typeset by Zack Taylor, www.zacktaylor.com

This book is printed on 100% post-consumer recycled, ancient forest friendly paper.

Printed in Canada

To Mary Ellen and Patrick

Contents

Lists of Figures, Maps, and Tables

Figures

Maps

Tables

Acronyms

ADt	Air-dried tonnes
AMR	AMR Corporation (parent company of American Airlines)
AOX	Absorbable organic halogens
ASA	Air Services Agreement
BCNI	Business Council on National Issues
BOD	Biological Oxygen Demand
CAC	Canadian Airports Council
CAI	Canadian Airlines International
CAL	Canadian Airlines Limited
CATA	Canadian Air Transport Association
CCCE	Canadian Council of Chief Executives
CCRA	Canada Customs and Revenue Agency
CEO	Chief Executive Officer
CEP	Communications, Energy and Paperworkers Union
CEPA	Canadian Environmental Protection Act
CITAC	Consuming Industries Trade Action Coalition
CITT	Canadian International Trade Tribunal
CMA	Canadian Manufacturers Association
CNR	Canadian National Railway
CPA	Canadian Pacific Airlines (CPAir from 1969-86)
CPPA	Canadian Pulp and Paper Association
CPR	Canadian Pacific Railway
CR	Concentration ratio
CRS	Computer Reservations System
CSPA	Canadian Steel Producers Association
CTC	Canadian Transport Commission
CTMP	Chemi-thermo-mechanical pulp

DDT   Dichloro-diphenyl-trichloroethane
DFF   Department of Fisheries and Forests (Canada)
DOE   Department of Environment (later Environment Canada)
DREE   Department of Regional Economic Expansion (Canada)
ECF   Elemental chlorine-free (pulp)
EMR   (Department of) Energy, Mines and Resources (Canada)
ENGO   Environmental Non-Governmental Organization
FSNC   Full service network carrier
FTA   (Canada-US) Free Trade Agreement (1989)
GATT   General Agreement on Tariffs and Trade
HR   HR product; hot-rolled sheet and coil steel
IPSCO   Interprovincial Steel Company
ITC   International Trade Commission (US)
kg/ADt   kilograms of AOX per air-dried metric tonne of pulp
LCC   Low-cost carrier
LTV   LTV Steel Corp.
NAIC   North American Industrial Classification
NAFTA   North American Free Trade Agreement
NSBK   Northern softwood bleached kraft
NSFI   Nova Scotia Forest Industries Ltd. (1971-85)
NSPL   Nova Scotia Pulp Ltd. (1958-71) (see also NSFI, SFI)
OECD   Organization for Economic Cooperation and Development
OPEC   Organization of Petroleum Exporting Countries
PPER   Pulp and Paper Effluent Regulations
PPMP   Pulp and Paper Modernization Program
PWA   Pacific Western Airlines
SFI   Stora Forest Industries Ltd. (1985-98)
SGF   Société générale de financement
SIMA   Special Import Measures Act
SWOC   Steel Workers Organizing Committee
TCA   Trans-Canada Airlines
TCF   Totally chlorine free (pulp)
TMP   Thermo-mechanical pulp
TRQ   Tariff rate quotas
TSS   Total suspended solids
US   United States of America
USWA   United Steelworkers of America
VRA   Voluntary Restraint Agreement
WTO   World Trade Organization

Preface

In the closing days of the year 2000, the *Globe and Mail* asked 30 Canadian corporate leaders about the challenges that would worry them over the next 12 months (*Globe and Mail* 2001). As might be expected, the responses were diverse. There were, however, persistent themes that permeated the answers regardless of product line or industry sector.

Only two chief executives cited competitors with whom they were locked in takeover or market rivalries as their prime concern. A far greater number (ten) identified market forces, expressed in financial, technology, or business cycle terms, as their greatest challenge. Almost an equal number (nine) cited factors arising from state policy, such as exchange rates, business confidence, regulatory controls, government budgeting, and the engineering of a "soft landing" at the close of the 1990s boom. However, the greatest number (13) of business leaders in the poll pointed to issues arising within their firms as the most crucial. These ranged from managing new acquisitions and reorganizing existing holdings to retaining trained staff and improving price/earning ratios.

In the context of this book, all four categories bear important political connotations. That one-third of these executives would point to state policy decisions as crucial to their futures gives the lie to arguments that government has become irrelevant in the age of global capitalism. Instead, it appears that the state remains a prime factor in the context of modern business management. Politics, as conventionally defined, still matters.

Yet the other thematic responses are equally revealing. In-house challenges are clearly on the minds of this pool of corporate leaders. They are preoccupied with strategic decision-making and organizational change within their companies. Clearly, the business firm is a complex amalgam of rivalrous and contradictory interests, which must be mobilized and nurtured toward desired objectives. In the chapters below, I argue that such internal relations of authority within the business firm are inherently "political" as well.

Finally, we should remember that one-third of the respondents pointed to market forces as their greatest concern. This is hardly surprising, since modern markets are dynamic and volatile, and they generate the short-term signals that drive corporate life. Factors such as the cost of capital, the share price of equity, the terms of international trade, and trends in business or industry cycles are irreducibly economic in character. Yet even here it is important to recognize the role of political forces as part of the equation. Interest rates are shaped closely by the monetary policies of central banks. The stock exchanges, where share trades take place, are regulated by state authorities. And the prospects for import and export trade are sensitive to the tariff and non-tariff policies that are delivered by governments within the framework of the World Trade Organization (WTO) and other agreements. Even in the neo-liberal environmental of global capitalism, it is evident that markets are constituted and maintained through political as well as economic channels.

In sum, the *Globe and Mail*'s expert panel offers valuable reminders about the real world of Canadian business and its underlying political dimensions. These unfold on a number of distinct but related planes, much as the artist portrays in the image found on page 2. Players are arranged differently on each level, and players can move between levels depending on circumstance, interest, and purpose. The micropolitics approach, developed in the chapters below, aims to improve both our awareness and our understanding of this world.

The Micropolitics Approach

The politics of Canadian business is attracting more analytic attention than ever before. This is reflected in historical studies of firms and corporate leaders and associational studies of business organizations and groups, as well as policy studies of the state interventions that shape market, industry, and company behaviour. For students of business politics, an impressive range of texts, edited collections, and monographs offer insights into the relations between market and state. These include neo-pluralist interpretations of interest group networks and policy communities, industrial organization frameworks based on models of market structure, corporate elite studies rooted in organizational hierarchy, and neo-marxist approaches to power by way of accumulation and class. Each of these adds important elements to our understanding of business-government relations.

At the same time, the past decade has seen an explosion of non-academic raw material involving business politics. This is evident, for example, in the dramatic expansion of policy advocacy work. Much of it has been sponsored by think tanks, consultants, and trade associations seeking to shape policy agendas to their own advantage. Equally, the rise of specialty television programming has brought business and public affairs into our homes and offices as never before. Third, the phenomenal expansion of traffic on the worldwide web makes mountains of "documentary" material accessible with unprecedented speed and scope to the interested public. In short, there is a bewildering jungle of data available to enthusiasts of business-government relations. The challenge is how to make sense of it.

It is here that I hope the micropolitics approach can play a constructive role. It offers a set of conceptual tools and models that can illuminate the politics of everyday business at the levels of industry, firm, and policy issue. It builds wider contexts in which the concrete particulars of business-government relations can be explored and understood in a systematic and comparative fashion. For students, it offers the opportunity to work with a variety of analytic materials that help answer the question "why?" Despite its many strengths, the literature of business politics still suffers from a serious shortfall in fine-grained, molecular studies. By molecular, I refer to a level where single industries and their constituent firms interact with state authorities as the prime objects of analysis. While this is constantly shaped by wider, macro-level forces, our overriding concern here is with specific political-economic actors and the communities of interest they embody.

The latter are critical, since so much of business politics springs from this "micro" level. Episodes surface informally each day in the business press, only to recede and disappear in the face of the next fast-breaking story. Although a file stuffed with press clippings is integral to this kind of work, it is insufficient in itself. First of all, it offers an imperfect record, due in part to the limitations of journalism in deciding what is newsworthy and in part to the reluctance of the principal protagonists—the managers, financiers, investors, politicians, and bureaucrats—to collude in drawing public attention to their affairs. Second, effective analysis depends on asking the right questions, and for this the public record offers limited clues at best. Yet it is important, both for students and citizens, to understand this powerful domain of social life. By exploring how analytic rigour may be brought to the study of inter-corporate as well as intra-corporate politics, and by illustrating some continuing applications, this book may make a modest contribution. There are literally hundreds of industries and businesses waiting for study.

What is the point of such comparisons? Taken together, they illustrate a range of possible variations in business-government relations. Case-by-case study may help determine where the peculiarities of individual firms or industries hold sway and where more enduring uniformities or structures prevail. It must be said, however, that a set of only three cases can be suggestive but never definitive. Comparison can also be viewed as a controlled exercise in factor analysis. If carefully planned, this permits an investigation of particular features of political economies by comparing deliberately chosen cases. Obviously, the selection of cases is critical. Normally, the list should be determined by the shared qualities of greatest interest, while the cases may in other respects differ radically.

Our choice here is shaped by our object of investigation, which is most broadly described as the internal structure of industry politics in contemporary Canada. As already mentioned, the three industries are pulp and paper, primary steel, and passenger air transport. This comparative set represents a revealing cross-section of a vast economic field. It samples from the primary (resource), secondary (manufacturing), and tertiary (service) sectors of the economy. It includes one industry (pulp and paper) that is primarily export-oriented, one (steel) that is focussed mainly on domestic sales, and one (air transport) that is strongly grounded in both.

In addition, each industry came to prominence as part of a different stage in the history of capitalism. Historical geography distinguishes a series of long (40-60 year) growth waves named after the Russian economist, Nicolai Kondratieff. Each Kondratieff wave is associated with a cluster of technology changes that generate persistent growth as they diffuse through economies by key "carrier" industries (Shuman and Rosenau 1972). For example, the second wave began in Europe in the 1840s and was based on steam power, rail transport, and coal energy. Primary steel was part of this cycle, driven by advances in metallurgy and coal-fired energy. Together they led to the mass production of high-strength materials for use in machinery and construction. The third Kondratiev wave was launched in the 1890s and continued through the early twentieth century. Pulp and paper is a good example of these new technology clusters, since it combines electricity and industrial chemicals to transform wood fibre into a variety of cellulose products. Beginning around World War II, a fourth wave began to build in association with petroleum-fuelled engines and synthetic materials. The postwar surge in civil air transport combined with advances in jet propulsion, aeronautic design, and computerized instrumentation to open a new era of accelerated global travel.

Arising in part from such differences, our three industries have generated unique constellations of political interests and singular strategies. Each indus-

Table P-1

Matrix of Industry Variations

	Pulp and Paper	Steel	Airlines
Primary Sector	○	○	
Secondary Sector	●	●	
Tertiary Sector			●
International Market-Oriented	●	○	●
Domestic Market-Oriented	○	●	●
2nd K: Iron / Steel Wave		●	
3rd K: Chemical / Electrical Wave	●	○	
4th K: Aeronautic / Electronics Wave	○		●
Tariff	●	●	
Non-Tariff Barriers		●	○
State Enterprise		●	●
Procurement		●	
Tax Expenditures	○	●	
Regulation	●		●
Competition Enforcement	○	○	●
Infrastructure			●

● Strong association
○ Lesser association

try has sought, and achieved, extended state support through policy interventions of widely differing types. They range from tariff levies and tax incentives to regulatory regimes and state enterprise. Clearly, there are a number of intriguing contrasts to probe. Some of the key variables are summarized in the matrix of industry variations in Table P-1.

Not everything about this trio of industries is distinguishing, however. There are also some common denominators. Each has been a Canadian success story in its own time. Pulp and paper ranks among the country's leading export manufacturing sectors, with major integrated facilities in every province except Prince Edward Island. Primary steel has been justly celebrated as an indigenously owned, highly innovative, and efficient supplier of commodity materials to the domestic market (Singer 1969). The passenger airline system is another example of Canadian-owned enterprise, providing safe and stable service to a far-flung nation that relies upon air transport for economic cohesion (Stevenson 1987).

In addition, each industry has experienced major technological revolutions over the course of its product cycle. The standard processes of paper production have shifted from mechanical to chemical pulping methods, while bleaching processes have switched from chlorine to oxygen treatments. Along the way, the speed of papermaking machines has accelerated dramatically along with the capacity to produce a spectrum of paper grades (Sinclair 1990). In steel, the open hearth blast furnace that ushered in mass production gave way, after World War II, to the basic oxygen furnace. Similarly, the advent of continuous casting brought cost savings and efficiencies that revolutionized industrial production. Later still, the design of electric arc furnaces opened the way for steel scrap to be recycled in mini-mill facilities of smaller capacity but significantly greater efficiencies (Ahlbrandt, *et al.* 1996). In its turn, air travel has witnessed several dramatic engineering transitions. The first was the transition from piston to jet aircraft, which meant a quantum leap in aircraft speed and capacity. The next was the emergence of a new generation of small commuter and regional aircraft, which made servicing secondary markets profitable (Doganis 1991).

More recently, each of the three industries has faced wrenching changes, most particularly by the global restructuring processes of the past two decades. Part of this involves the rise of major competitors from other parts of the world. Fast-growing tropical timber has drawn a significant share of market pulp production from North America and Scandinavia to South America and Asia (Marchak 1995). The new steel capacity that has appeared since World War II, particularly in Japan and South Korea, has made inroads in world steel trade. Meanwhile, the lingering overcapacity in western and eastern Europe releases low-priced exports of a different sort (Hogan 1983). Airlines represent a partial exception, as state-to-state commercial air treaties continue to dictate the shape of international business while significantly dampening the competitive dynamic (Hanlon 1996).

Today, all three sectors remain vital if problematic blocs in Canada's economy. In part because of this, each faces challenges of the first order on the political front. One continuing pressure is for effective political organization, both at the industry-wide and at the company level (Coleman 1988). No matter how intense their commercial rivalries, firms may acknowledge a shared set of interests when it comes to relations with government—or with other industries. The industries examined here have fashioned a variety of trade associations to mobilize their common interests. At times, this has facilitated the effective exercise of collective pressure, while at other times, intra-industry commercial tensions have triggered political rivalries as well. The Canadian Pulp and Paper Association (now the Forest Products Association of Canada) is one of the oldest and largest such organizations in the nation. Dating back almost a century, it has played a crucial role in organizing an industry consensus on matters of engineering, taxation, and trade (Grant 1990). By contrast, the Canadian Steel Producers Association was founded only in the 1980s, in the midst of a protracted series of trade disputes with US producers (Kymlicka 1987). Prior to that time, the major Canadian firms operated through the American Iron and Steel Institute. For airlines, the Air Transport Association of Canada played a prominent representational role in the regulated era from 1937 to 1984. However, its strength was sapped by the domestic corporate takeover and merger wave of the latter 1980s. Once the major commercial rivals were reduced to two, their intense struggle saw the axis of government relations resolve to the firm level.

Ways to Use this Book

In the chapters below, the three industries are probed in depth. But before tackling the detailed cases, an introductory chapter outlines a series of concepts and models that will guide the investigations by offering some tools that can be applied to the exploration of particular firms, sectors, and public policies.

Readers are encouraged to approach this book in an interactive fashion. There are several possible strategies for reading the text. For example, it is possible to "read through" each chapter in a linear sequence in search of a rounded perspective on the industry in question. Chapters 2, 3, and 4 display a certain symmetry, since each industry is presented in a similar way. They begin with comments on the contemporary (millennium) setting and then proceed to explore both the historical background and the structural underpinnings of industry activity. Then, to sample the internal politics of firms, we examine a pair of leading corporations in each chapter. To illustrate some dynamics of inter-firm and industry politics, we explore two significant policy issues arising in each industry. Finally, each chapter closes with a glossary of key terms and relationships. Overall, the challenge of this book is to grasp how causal relations cascade from one dimension of an industry, or chapter, to another. Keep in mind, however, that industry politics is anything but homogeneous. Within each chapter, the firms and policy cases have been selected to illustrate variation as much as uniformity. Avoid simple generalizations. When in doubt, opt for comparison and contrast.

At the same time, it is also possible to read "across" industries or chapters by selecting a single theme or topic and assessing the variations from one industry to the next. Make an active attempt to compare the situations in the three industries, searching for the roots of similarity and difference. What separates the experiences of the paper, steel, and airline businesses in the

Table P-2

Comparative Dimensions of Three Industries

Pulp and Paper	HISTORY
STRUCTURE	P/P Effluent Domtar · Stora P/P Modernization

Air Transport	HISTORY
STRUCTURE	Open Skies Air Canada · CP Air CAI ONEX- AC Merger

Primary Steel	HISTORY
STRUCTURE	Algoma / Sysco Rails Stelco · IPSCO Safeguards

year 2000? What are the key historic forces and turning points shaping the developmental path followed by each of the three sectors? Does industry structure impinge differently on corporate strategy from one firm and sector to another? Which of the many aspects of inter-firm politics are manifest in each corporate profile? What variations of policy network, decision cycle, bureaucratic politics, or issue management are decisive in shaping a public policy outcome? Table P-2 illustrates the structure of the industry chapters and some possible pathways for comparative study.

This is what is meant by an interactive approach to the text. Move freely within a chapter in the search for rounded understanding. Jump from a section in the frameworks chapter to the corresponding parts of the industry chapters, and back again. Frame your own cross-chapter comparisons. Finally, remember that the tools and perspectives found in the chapters below can also be applied to the study of other industries. Hopefully, readers will be encouraged to bring new sectors under the microscope as well.

Acknowledgements

The idea for this study occurred almost a decade ago, during travel in Europe. More specifically, it was inspired by one day's reading of the *Financial Times*, which for my money is the world's best newspaper source on the micropolitics of business. That day, the coverage was particularly rich on the involvement of paper, steel, and airline interests in shaping state policy. While Canada does not normally figure strongly in the pages of the *Financial Times*, the parallels were striking, and the seeds for this project were sown.

There have been many iterations over the intervening years. I wish to thank my students of business and government at St. Francis Xavier University, who heard these ideas first. They proved to be a valuable sounding board, and the arguments were much improved by their reactions. Work on the manuscript also benefited in major ways from my access to the following collections: the British Library of Science and Technology (now housed in the British Library) in London; the University Library of the University of Cambridge, England; and the DalTech Library in Halifax, Nova Scotia.

Sources

Image of Bowler-hatted Capitalist
Reproduced by permission of the artist, Mr. Eugen Mihaescu.

Figure 1-3: Decision Flow Chart for Firm Rationalization
Reproduced by permission of Oxford University Press Australia from *Turning It Around: Closure and Revitalization in New Zealand Industry* by Savage and Bollard © Oxford University Press. www.oup.com.au

Map 2-1: Location of Pulp and Paper Mills in Canada, 1988
Source: Controlling Pollution from Canadian Pulp and Paper Manufacturers: A Federal Perspective—Map 1 (7). *Location of Pulp and Paper Mills in Canada*, Environment Canada, 1990. Reproduced with the permission of the Minister of Public Works and Government Services Canada, 2004.

Figure 3-1 A Flow Line for Steelmaking
Reprinted with permission from *World Steel Statistics Monthly*, published by ISSB Ltd, London, England.

Figure 3-4 Anti-Dumping and Countervail Process
Source: "What You Should Know About Dumping and Subsidy Investigations." Canada Customs and Revenue Agency. Reproduced with permission of the Minister of Public Works and Government Services Canada, 2004.

CHAPTER ONE

Concepts and Frameworks

Truly there are times when a picture seems worth a thousand words. Consider the illustration by Eugen Mihaescu (on page 2). This image of the portly but faceless capitalist, replete with bowler hat, greatcoat, and walking stick, was published in 1971 in the *New York Times*. But it is the activity in the foreground that draws our attention, for it is here that the archetypal magnate unfolds literally into a series of layered platforms, where smaller capitalists can be found. Some appear in isolation, while others are clustered in groups. This image captures, in a simple yet compelling way, the micropolitical world of Canadian business explored in the pages below. Big capital, embodied in the man with the bowler, opens out to reveal a series of inner levels or stages on which different configurations hold sway. As with the industries and sub-industries in this book, each level signifies a field of activity with its own characteristic features even as it remains tied to the larger whole. Only one thing is missing—the dynamic forces that animate the respective players. In this chapter, we consider a series of possible engines that drive industry politics.

Politics is a creative, open-ended activity capable of surprising variation, but it is not conducted on a blank page. Indeed, much of its practical art and analytic science involves an understanding of the conditioning patterns that lurk below the surface. In the study of social life, this duality is often presented as a tension between agency and structure. Human beings, individually and in groups, are certainly capable of independent action of their own design and practice. They are in this sense "free agents," whose reflexive action is framed by conscious interests, objectives, and plans. Yet humans are at the same time located within fixed societal fields or structures. A variety of continuing structures impose constraints on social action across both space and time. It is in this sense that politics should be understood as an open-ended pursuit, but one still subject to deeply embedded force fields that must be reckoned with.

Figure 1-1

Actors and Structures

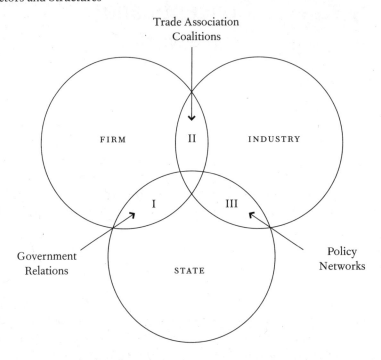

This chapter introduces some perspectives on the structural side of micropolitics. Our aim can be captured in a simple model, which highlights how politics is generated in each of three central domains. Figure 1-1 sets out the key relationships. It begins with the internal politics of the *firm*, where key strategic decisions originate. This is the level where the most specific and unique characteristics emerge and where each business unit defines its particular political interests and goals. A second domain is the aggregate politics of the *industry*, where groups of firms can set aside their commercial rivalries to articulate common political concerns, particularly in the face of external political challenges. Though there are many barriers to effective industry organization, group political action is integral to micropolitics. Third, it is important to explore the mechanics of *state* authority. Here public policy measures, in the form of laws and programs, can alter—sometimes decisively—the contours of business.

It is also interesting to note that there are several overlapping segments among these three domains, which represent fields of encounter where specific political processes hold sway. In Field I, for example, firms may interface directly with state authorities on issues of particular concern, such as tensions among rival companies who are competing for state policy outputs. This is especially pertinent to procurement and regulatory licensing issues, which tend to assume zero-sum proportions — one company's success means other companies' failures. In Field II, however, firms need to determine the extent of their shared interests and the modalities to advance these collectively. Trade associations and business coalitions are prime instruments here. Finally, in Field III, industries need to establish ongoing relationships with government bodies that are in a position to significantly shape their operating environments. These can be usefully captured in the notion of policy networks.

Exploring the Politics of Firms

We begin this discussion with the individual firm, the most basic type of business organization which exists to make money. In moving to this most concrete level of analysis, we open a new and equally complex perspective on business politics. While each firm shares certain characteristics of the industry as a whole, it also competes with its rivals in search of maximum returns. To this extent the interests of each firm are at least partly in conflict with all others. In order to realize its goals, a firm must take decisions and pursue strategies aimed at self-maximization. This market rivalry can easily carry over to the pursuit of competitive advantage in the political realm as well.

But what drives these decisions and strategies? There is a common but mistaken tendency to see more consistency of purpose and continuity of interest within the business firm than in other complex organizations such as governments, churches, universities, or trade unions. Where the latter seem to bristle with internal tensions and competing agendas, the firm is assumed to be imbued with discipline and unity. Perhaps this is reinforced by economic models that adopt highly simplified views of firms as single-interest maximizers. Such unity of purpose might be achievable in the smallest family-based proprietorships and partnerships — though even this may be suspect — but it certainly does not always fit the world of medium-sized and big business.

The Firm as a Political Coalition

In a far-sighted but much-neglected article published in 1962, James March observed that "the firm in economic theory is more commonly treated as the basic unit in a larger conflict system (industry, market, economic system) than as a conflict system in itself" (March 1962, 668). Yet it is precisely this inner conflict system that confers a political character on the firm, and the more diverse the interests, the more complex the power relations within. March went on to advance an alternative view that highlights the political processes within the firm. He proposed that "we assume that a business firm is a political coalition and that the executive in the firm is a political broker. The composition of the firm is not given, it is negotiated. The goals of the firm are not given, they are bargained" (March 1962, 672).

This presents the internal constituents of the firm in a new light. Factors of production such as land, labour, and capital are not simply "available" but must be martialled and combined in particular ways. March continues:

> We assume that there is a set of potential participants in the firm. At least initially, we think of such classes of potential participants as investors (stockholders), suppliers, customers, governmental agents, and various types of employees. More realistically, we might supplement such a list with such actual or potential participants as investment analysts, trade associations, political parties and labor unions. Each potential participant makes demands upon the system. (March 1962, 672-73)

Notice that even the smallest business faces most of these issues. Even if labour is drawn from family members (and this does not always apply today), there is the question of obtaining capital, facilities, and supplies; developing a clientele; paying taxes; and so on. Clearly, there is potential for political relationships to shape the daily operations of any type of firm. At the same time, March points out that a firm manages these complex challenges through its own internal authority system. As a broker, the business executive is expected to field these challenges and to reconcile the cross-cutting pressures through management and planning. However, his or her job is no less "political" for the fact that it unfolds within a corporate office.

A fascinating illustration of the internal politics of today's mega-firm is offered by David Kearns, in his business memoir *Prophets in the Dark*. In 1982, Kearns gained the top job at Xerox Inc., a truly massive corporation, with more than 100,000 employees worldwide, ranking thirty-eighth on the Fortune 500 list, and earning in excess of $1b the previous year. Yet, rather than taking solace in such evident strength, Kearns reports that he was terrified for his firm's future. He saw superior products emerging from Japan, unprecedented turbulence in Xerox's global markets, and signs of declining market shares. As president and chief executive officer (CEO), he felt bound to keep this apprehension to himself. If the news spread within the firm, he risked the departure of key staff and, if it became public, it would jeopardize Xerox stock prices, credit ratings, and customer sales. Working privately, he developed a two-prong strategy: diversify production in the short run and remake Xerox into a high-quality company in the long run. The first goal was achieved in relatively short order, with the support of the board of directors and outside investment bankers. But the second required almost a decade of sustained effort. As Kearns described it:

> I'm trying to get a hundred thousand people to act and think differently toward the product, the customer, and each other every hour of every day. All of my senior managers say they agree with me. But I feel that if I leave the company for a couple of days, they will go right back to doing what they were doing. (Kearns and Nagler 1992)

This is a classic diagnosis of a political situation, in which the top manager faces inertia and self-interest in the bid to restructure. Kearns was an early US convert to the doctrine of Total Quality Management as the basis of a new Xerox operating strategy. In pursuit of this, he adopted a series of classical political tactics: enlisting external consultants as allies; converting or neutralizing the balance of his senior management team, described as the six "kings" and eight "princes"; taking personal sponsorship of the quality program; widening the range of engaged interests by holding special training sessions for all employees worldwide; and managing the selection of his successor so as to secure any gains that he made.

The Xerox story is more typical than unique in this respect. While strategic goals may differ, most company heads face a formidable set of internal challenges when it comes to implementation. We can now consider a general model of the modern corporation from the point of view of its constituent interests, as set out in Figure 1-2 below.

Note that there are at least two levels of political relationship. First, within the corporation there are legal relations of ownership and control, linking shareholders, directors, and managers. These are often designated as the relations of "corporate governance," a telling phrase that underlines once again the political character of intra-firm decision-making. While shareholders figure as the legal owners, it has long been recognized that only very large block holders are able to translate that clout into action. For most sharehold-

Figure 1-2

The Internal Politics of the Firm

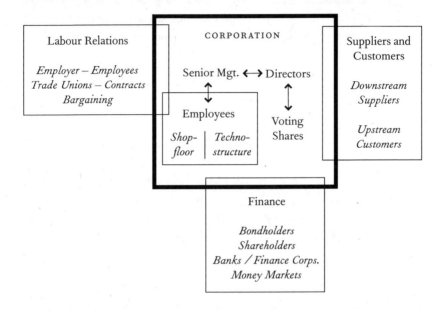

ers, involvement starts and ends with a mail-out announcing the corporate annual meeting, when yearly dividends are announced, the annual report is released, and directors are elected to the company board. In firms with widely dispersed shareholdings, *de facto* control tends to be shared between the board and the senior professional managers whom the board is charged to oversee. In practice this relation too is subject to much variation. The textbook distinction between major policy decision-makers (boards of directors) and day-to-day implementers (senior executives) quickly dissolves. Instead we encounter a universe of "activist" (hands-on) and "passive" (deferential) boards, variable balances of "inside" (managerial) and "outside" (elsewhere employed) directors, cases of widely dispersed (non-controlling) shareholdings resulting in "management control," and cases of concentrated holdings by "majority" or "controlling" blocks.

To the extent that these variations shape decision-making, they will have a major impact on the internal politics of the firm. Peter Cook's historical portrait of the Massey-Ferguson Company offers dramatic testimony here. Once Canada's most powerful multinational producer of agricultural machinery, with subsidiaries spanning the globe, Massey collapsed into near-bankruptcy in the late 1970s and the name almost disappeared without a trace. Cook illustrates the connection between ownership, management, and industrial decline (Cook 1981).

The company began in the nineteenth century, when patriarch Daniel Massey transformed his blacksmith business into manufacturing farm implements. Three generations of family members led the firm over its first half-century. However, the Massey family lost control during the 1920s after the firm was floated on the Toronto Stock Exchange, and a period of management control followed. Then, following World War II, the Argus trust led by Toronto financiers J.A. (Bud) McDougald and E.P. Taylor acquired a controlling block of Massey-Ferguson shares. Argus maintained tight policy control over each of its operating companies by dominating the executive committees of their boards of directors. In this way, Argus closely shaped key financial decisions on dividend remits, retained capital, and corporate debt. Argus expected to maximize its earnings from each holding within the group, a priority that did not always fit the strategic needs of the operating firms. Massey was left with paltry retained earnings in the critical period of postwar expansion and was forced into heavy corporate borrowing to obtain investment capital. By the 1970s, a combination of record high interest rates on this weighty debt load and a deep cyclical slump in demands for farm machinery left Massey in a far more desperate condition than its competitors. It is important to note how much these troubles were particular to the firm and its conglomerate power structure. In the very year that Massey-Ferguson foundered, its rival John Deere was honoured as one of the best-managed firms in North America.

For the past generation, it has been common to distinguish two groups of paid employees below the senior echelon: the salaried (white-collar) middle managers and the waged (blue-collar) shop-floor workers. The typical points of contrast are well known: the higher income middle managers identify more closely with corporate interests than do the wage workers, who often look to trade unions to protect their interests in contract bargaining with corporate employers. In an influential though somewhat dated study, J.K. Galbraith suggested that middle management was the indispensable control mechanism within the modern firm. Galbraith designated this the "technostructure," the expression of organized intelligence that made sophisticated corporate planning possible on an unprecedented scale (Galbraith 1978). Indeed, he went so far as to suggest that senior managers and directors had lost their earlier pre-eminence. Galbraith argued that the technostructure was now the driving force of the large corporation and that the triumph of the planning impulse meant that growth had supplanted profitability as the prime objective. While steady dividends were required to placate shareholders, and reasonable profits were necessary to assuage banks and lenders, he suggested that the drive to increased scale of business was now fundamental to the largest firms which together formed the "planning system." In Canada, the apex of this system is occupied by the sort of firms listed in Table 1-1.

It is not necessary to resolve this sometimes strenuous debate to appreciate Galbraith's contribution to the political analysis of the firm. Furthermore, his distinction between the "competitive" and "planned" sectors of the corporate world suggests separate dynamics for each. The notion that the logic of planning has captured both big business and big government implies certain possibilities for mutuality and cooperation between them. Finally, the bifurcation of the workforce carries important implications. The presence of a trade union introduces an important counterweight to executive control of labour costs, usually a key factor in the total cost of production. This is further reinforced by the labour relations regimes established in state policy, which both regulate and legitimize the organization, certification, and contract bargaining by unions. While these regimes have come under steady assault in recent decades, organized labour continues to play a crucial political role in all three industries examined in this book.

Another externally constrained factor of production is the supply of investment capital. Indeed, if labour relations constituted a volatile factor in the 1960s and 1970s, given tight labour markets and aggressive unions, then capital assumed this status for the 1980s and 1990s. This is evocatively captured in the term "casino capitalism" (Strange 1986). Huge pools of capital had accrued in the form of petro-dollars, following the OPEC (Organization of Petroleum Exporting Countries) oil shocks, and super-profits from criminal trade or so-called "hot money" (Naylor 1987) and money market trading accounts, in the era of floating exchange rates and volatile interest rates. Once the new communication technologies permitted worldwide trading, instant electronic sales, and computer-programmed transactions, the financial markets came to resemble a vast casino with "bets" being wagered through a variety of high-risk but lucrative instruments (Henwood 1998). By the mid-1980s, the superior returns from financial trading threatened to eclipse equity trading and directly productive investment as preferred outlets for capital. By the 1990s, the machinations of global money markets could trigger currency collapses, such as those in Asia in 1997 and in Russia in 1998, and shake the foundations of governments, such as in the United Kingdom in 1992 and in Argentina in 2002.

This had dramatic effects on the politics of the firm. Corporate finance leapt from one of the most stable and prosaic of business disciplines to one of the most dynamic and explosive. One emblem of big business in the 1980s was

Table 1-1

Top Canadian Companies, 2002

Profits, 2002			Revenues, 2002		
1	Canadian Wheat Board	$4.272 b	1	General Motors of Can.	37.000 b
2	Caisse de dépôts et placement	3.469 b	2	George Weston Ltd.	27.446 b
3	Royal Bank of Canada	2.762 b	3	Bombardier Inc.	23.664 b
4	BCE Inc.	2.475 b	4	Ford Motor Co. of Can.	23.328 b
5	Ontario Lottery Corp.	1.979 b	5	Royal Bank of Canada	23.234 b
6	Bank of Canada	1.822 b	6	Sun Life Financial	23.101 b
7	Bank of Nova Scotia	1.797 b	7	Onex Corp.	22.653 b
8	Alberta Gaming/Liquor Comm.	1.612 b	8	Magna International Inc.	20.364 b
9	Hydro-Québec	1.526 b	9	BCE Inc.	19.768 b
10	Loto-Québec	1.448 b	10	Alcan Inc.	19.687 b
11	Bank of Montreal	1.417 b	11	Daimler Chrysler Canada	19.353 b
12	Manulife Financial Corp.	1.370 b	12	Power Corp of Canada	19.017 b
13	Encana Corp	1.224 b	13	Bank of Nova Scotia	18.310 b
14	Imperial Oil	1.210 b	14	Can. Imp. Bank of Comm.	17.055 b
15	Sun Life Financial Services	997 m	15	Imperial Oil Ltd.	16.890 b
16	PetroCanada	977 m	16	Toronto Dominion Bank	16.680 b
17	Thomson Corp.	966 m	17	Nortel Networks Corp.	16.538 b
18	Liquor Control Board Ont.	921 m	18	Manulife Financial Corp.	16.532 b
19	Magna International Inc.	870 m	19	Bank of Montreal	13.059 b
20	Mouve. des Caisses Desjardins	848 m	20	Hydro-Québec	13.002 b
	…			…	
85	Domtar Inc.	141 m	27	Air Canada	9.826 b
303	IPSCO Inc.	31.8 m	47	Domtar Inc.	5.490 b
303	Stelco Inc.	14 m	98	Stelco Inc.	2.784 b
565	Air Canada	− 428 m	159	IPSCO Inc.	1.698 b

Source: *National Post Business* 2003. ($Can.)

the technique of the leveraged buy-out, which well illustrates the transformative potential, under certain conditions, of short-term speculative finance (Bruck 1989). At a sufficiently high interest-rate premium, capital could be raised in the form of "junk bonds" for firms whose business fundamentals would not normally justify the investor risk (Madrick 1996). The issuers of such securities would then apply the proceeds to corporate takeover campaigns that targeted other firms whose share prices lagged below their real asset value. Following a successful takeover, the captured firm could be broken up, resold, and reconfigured, often at extraordinary profits to the "raiders."

As the frequency and scale of these transactions mounted, even the largest international firms became vulnerable. For potential targets, the management of share prices became an essential element in corporate self-defence, together with "poison pill" provisions. The latter were written into the terms of corporate governance to discourage raiders by making the challenge too complex or expensive a proposition. For example, a hostile (i.e., uninvited) takeover bid could trigger an issue of new shares or warrants (the right to buy shares) to existing shareholders in the target firm. This increased both the cost and the difficulty of the prospective takeover. Shareholder rights schemes played a similar role, by obligating potential buyers to offer the same terms to all shareholders, rather than restricting tenders to a few large shareholders conveying majority control. Even unsuccessful takeover bids could be immensely lucrative to corporate raiders and were frequently mounted with this express purpose. Known as the "greenmail" tactic, this involved announcing a formal takeover bid and accumulating all available shares in the expectation that the trading value of the stock would rise both as a function of the tender itself and from defensive purchasing of shares by the target firm. The raider then sold the accumulated shares in the speculative market, taking profits on the difference between the initial and the run-up prices. Another available tactic under such conditions was the "friendly" merger or takeover, a pre-emptive move launched by a potential target firm to guarantee its future integrity under more acceptable conditions.

Strategic manoeuvring around questions of ownership, control, growth, merger, and acquisition touches all of the leading corporations profiled in this study. In pulp and paper, Domtar evolved under the direction of a parent holding company from a tar and chemical producer into one of eastern Canada's leading paper manufacturers. This continued after control shifted to the Quebec government-owned pension fund, the Caisse de dépôt et placement du Québec. Stora Forest Industries operated in quite a different organizational setting, as a wholly owned Nova Scotia subsidiary of one of Scandinavia's largest forest-sector giants. In steel, industry leader Stelco has long been owned by a diversified shareholder base, giving senior management a high degree of policy control. Until the early 1980s, the Steel Company of Canada (Stelco) was regarded as a solid blue-chip investment. However in the years since, and particularly since 1992, the company has been locked in an ongoing struggle for survival. Share prices have oscillated wildly, institutional investors have become actively engaged, and senior management was ousted in a dramatic corporate turnover in 2003. The Interprovincial Steel Company (IPSCO), a mini-mill operator, has followed almost the opposite path, rising from modest roots as a prairie tubular pipe manufacturer to a highly profitable multinational producer of a wide spectrum of steel goods. IPSCO has reported profits every year since 1990 and has become a favoured equity bet on both sides of the border. Along the way, it has combined new plant investments with acquisitions to create today's asset network. Finally, the passenger air sector offers a panorama of ownership and restructuring initiatives. Not only has the entire air sector functioned under legislated limits for minority foreign equity ownership, but several leading carriers were established by or taken into state ownership and later privatized in the 1980s. This was followed by the epic, decade-long struggle between the two leading carriers for market dominance, featuring several failed merger and takeover efforts and ending with the Air Canada/Onex battle of 1999.

Without question, the casino economy revolutionized the power of finance capital in the corporate world. Certainly this unprecedented level of merger and acquisition activity transformed approaches to corporate strategy in the 1980s and reduced the Galbraithian theory of the impregnable corporate giant to historical proportions. At the same time, it renewed the debate over contending interests within the firm. Whose interests are served by a friendly or hostile takeover or a greenmail campaign? Are the institutional shareholders, who stand to benefit most from share purchase premiums, given privileged treatment over smaller shareholders, who are often ignored or treated separately? Are directors able to defend the interests of "their" firm during a takeover battle, without a veil of self-interest stemming from potential loss of directorial position or potential gain from share sales? Does the short-term investor focus on share-to-asset ratio and liquidation value play a positive role in sweeping inefficient business out of the market, or does it elevate short-term commercial profit-taking over a longer-term enhancement of productive assets? Finally, does the short-term, speculative focus of the casino economy serve the general economic or national interest, and what role should be played by state authorities (Hutton 1996; Soros 2000)?

Another set of interests is located at the periphery of the corporate organization, close to the heart of its operations. This involves client firms and customers positioned both upstream and downstream from a firm's core businesses. It may entail material suppliers, shippers, equipment manufacturers, or purchasers of output, whether for further value-adding or for final demand consumption. All of these elements represent interests subject to negotiation, with the index of priority hinging upon the indispensability of the linkage. A prominent supplier of a scarce commodity input, whether due to restricted production or excess demand, has the capacity to affect the firm's production function, as does a dominant supplier of transportation, energy, or communication services. One classic response to such situations sees the firm expand its scope of operations by integrating backwards along the production chain. The pulp and paper manufacturer may assume direct control through ownership or lease of the standing forest resource to ensure a steady long-run supply of logs. It may also develop a private hydroelectric or wood-waste energy supply to drive the plant. Similarly, the primary steelmaker may ensure its supplies of iron ore, scrap metal, coal, or bulk transport carriers by incorporating such operations into the enterprise rather than risking the vagaries of open-market procurement. Finally, the air transport carrier must control aircraft in order to sell passenger seats or freight space, but it also requires a sales system to connect with consumers, a maintenance system for equipment upkeep, and ground facilities for origin and destination services.

It is interesting that Galbraith stresses this enclosure of complementary activities through vertical integration as a powerful example of the planning system in action. Yet there is no inherent necessity for such intra-firm ties. True, they were common during the consolidating phases of North American industrialization (1890-1965). However, there are increasing exceptions to these rules: steel companies purchase market coal or scrap metal on arm's-length contracts; pulp companies buy fibre from private suppliers and on-spot markets; airlines lease rather than own their planes and out-source their maintenance or booking services.

Today it has been argued that the classic type of within-firm integration typified one historic phase of capitalism. Known as "Fordism," this model of industrial production was pioneered in the Ford Motor Company after World War I but permeated a broad set of modern, capital intensive industries after 1945. Its foundation was the mass production system that generated standard commodity outputs for stable consumer markets. It has been argued convincingly, however, that Fordism extended well beyond the shop floor to include a political regime of state fiscal management and labour relations (Harvey 1989; McBride 2001). In the decades since 1970, leading business sectors have been increasingly restructured around new production principles. These are flexible rather than rigid in their organization of material supply, labour, manufacturing processes, and outputs. Production centres on core sites of competitive advantage, with ancillary activities obtained by competitive contracting.

This is the "post-Fordist" world of flexible manufacturing, which adjusts for diverse product specifications according to customer; just-in-time delivery, which eschews high carrying costs for inventory in favour of tightly scheduled external supply; and lean manufacturing, which achieves cost efficiencies undreamt of in the earlier era (Harrison 1994). It also calls for a new policy regime to maximize the freedom of core business enterprises in the pursuit of profit opportunities.

Post-Fordist pressures can be seen in all three of our featured industries. For example, contemporary steelmaking technologies involving electric furnaces are fed by scrap metal that typically has not been backward-linked. Neither do these "mini-mill" steelmakers attempt to cover the full range of fabricated products; instead, they occupy niche markets best suited to their competitive advantages. Similarly, an entire segment of pulp and paper firms utilize waste fibre input derived from logging and sawmill by-products such as sawdust, slabs, or wood chips that now sell as bulk commodities. Equally, the air transport sector has seen a dramatic shift toward the utilization of small but operationally efficient aircraft to service non-trunk routes, as well as feeding passengers into company hubs for ongoing travel. Many experiments in non-reserved ticketing and walk-on/walk-off budget service are also efforts to gain price advantages.

Finally, it is important to note the role of state authorities as provisioners, as consumers, and as regulators of goods and services with private firms. This may take a variety of forms. Possible backward linkages include the lease or sale of state resources—forests, minerals, water power rights—and public investment in infrastructure—harbours and waterways, railways and highways, air navigation, and terminal facilities. It should be noted that state involvement may be achieved by market disposition (direct sale) or through state-owned enterprise. Forward-linked state consumption may play an equally important role, as reflected by the vast range of state procurement activities. The state figures prominently as a buyer of military hardware, communications services, computer hardware and software, building real estate, and consultancy services. Indeed, strategic state procurement has emerged as a crucial component of modern industrial policy. Also notable in the field of regulatory intervention is the manner in which the structures of markets can be mediated for political advantage. Small commodity producers, such as farmers, fishers, and woodlot owners, have frequently sought the support of state-sanctioned collective marketing arrangements, especially when they deal with tight buyer oligopolies. Here the principal goal is to increase the returns to the primary producers by conferring enhanced bargaining leverage, which of course has the effect of increasing input costs to upstream buyers. At the same time, states have also been pressured to regulate the pricing and service policies of quasi-monopolies in order to protect both industrial and domestic consumers facing imperfect markets. This is particularly common in infrastructural fields such as electricity production, transportation, and communication. It is revealing that seven of the top 20 Canadian companies, measured by profits, in 2002 were state enterprises, while another, Caisses Desjardins, was a credit union federation which flourished in a regulated environment (see Table 1-1).

In sum, our model of the firm as a political coalition reveals a series of intersecting interests, at both the centre and the margin of corporate life, which can directly shape the key outcomes by which business is measured. As March suggests, each firm can be seen as a negotiated structure, which imposes both opportunities and constraints on decisions that must be taken and allocations that must be made. Of course none of these are predetermined. Though many decisions are closely circumscribed, there is always room for choice, and a range of outcomes is possible. However, by basing our analysis on a realistic model of autonomy and constraint, we can avoid the analytic traps of over-determinism and over-voluntarism. In the chapters below we will explore the causal factors that have shaped particular outcomes at particular firms. For this, the concept of corporate strategy will be helpful.

Corporate Strategy

Later in this chapter we will see that a combination of industry categories and concentration ratios offers useful insights into market structure. It provides the basis for official statistical compilation for government reporting, international comparison, and the analysis of market structure. Useful as these categories and coefficients can be, they do not always disaggregate business relationships to the desired level. With industries being defined in terms of product or commodity lines, it is possible to miss the point that *firms* rather than *products* are the active agents of economic and political life (Eden 1991). Furthermore, firms are commonly involved in multiple product lines, sometimes involving extraordinary variation. As the calculus of decision-making becomes more complex, there is increasing need for strategic policy within the firm.

With his path-breaking work on the historical evolution of the business firm, Alfred Chandler injected an important analytic dimension. He drew particular attention to the early instances of diversified, multi-divisional management in such pioneering firms as Dupont, General Motors, Standard Oil, and Sears Roebuck (Chandler 1962). In contrast to Adam Smith's "invisible hand" of decentred market forces controlled by all and by none, Chandler argued that in the large modern firms that came to dominate many sectors "the visible hand of management replaced the invisible hand of market forces" (Chandler 1977). It is through the visible hand that corporate strategy unfolds: in the deliberate attempts by the constituent interests within the firm to define agendas, build alliances, take decisions, and realize the results.

Here the firm emerges once again as a political system in its own right. At one level, shareholders, directors, and senior managers vie to install their interests at the centre of corporate strategy. At another level, the managers, generally with the support of owners, join the ongoing battle with their workforces over conditions of employment and the distribution of revenue within the firm. At still another level, the different administrative branches of the firm engage in research, production, marketing, and finance, and manoeuvre to maximize their respective leverage in decision-making. This amounts to a distinct perspective on the large firm, and on business decision-making, in the oligopolistic sector. As Ann Markusen puts it:

> business strategy [must be cast] in the lead role, assuming that managerial decision-making is not perfectly dictated by the market but that risk-taking, organization-building, and market-dominating efforts, political influence, and mistakes may all be important contributors ... (Markusen 1985, 2)

Obviously, corporate political interests are integral to strategy. This begins with the acknowledgement that non-market forces can be critical to business prospects: they can neither be ignored nor taken for granted. Rather, the management of "political advantage" has become a significant subtheme of strategic management through the practice of government relations. Whether vested within the firm as a staff unit or contracted from outside consultants, it begins with the identification of corporate sensitivities and the forward scanning of the public policy environment. This points toward the tactical side of political intervention: deciding when to intervene, when to act alone and when to act in concert, and how to articulate policy concerns to greatest effect (Gollner 1983; Sawatsky 1987; Stanbury 1992).

Clearly this involves an interplay of both structure and agency factors. Certain of the former are inherent to the industries or subindustries in question. Are they open or closed to trade? Are they labour-sensitive, materials-dependent, or complex-factor (technology) based? Where are they located on the product cycle? At the same time, each enterprise within a market faces its own specific constraints in terms of governance, finance, profitability, and underlying growth potential, which must be reconciled ultimately within corporate circles. This agency focus helps to introduce an important dynamic component as a firm is seen as a product of past decisions and results.

One helpful illustration of strategic choice, which highlights the cascading nature of these qualities, is offered by Savage and Bollard (1990). Initially conceived to capture the pressures faced and options available to firms under strain from failing profits, it carries a broader relevance in our present context. A range of strategic responses is identified, ranging from "relocation" to "continuance" (long-term and short-term) to "mothballing," "revitalization," and "closure." As may be seen in Figure 1-3, outcomes are presented as branches of a decision-tree which turn on the relation of corporate earnings to production costs. In our adaptation, each point of decision is designated as a "step" whose outcome frames the remaining available options.

The real contexts of these somewhat abstract choices can be readily discerned in recent Canadian experience. For instance, Step 1 hinges upon the possibility of a firm being able to earn higher returns by shifting productive operations to a new location, a choice commonly faced when limits of scale, technology, factor cost, or shifting geography of consumption arise. If the answer is affirmative, then relocation ensues. Such questions have become increasingly urgent over the past 15 years, as post-Fordist realities have mounted in Canada. Significantly, state macro-policy can directly affect the process in such forms as deflationary policy (interest rates), capital markets liberalization (financial speculation), or trade policy (the Free Trade Agreement and the North American Free Trade Agreement). Indeed, this was exactly the calculation followed by the Kenworth Corporation in the spring of 1996 when it closed its truck plant in Ste. Thérèse, Quebec while expanding production in Tijuana, Mexico. This pattern has since been repeated many times.

Step 2 arises in the absence of superior location and turns upon whether revenue exceeds total costs of production. If the answer is affirmative, then business will continue along prevailing lines; if negative, the firm faces further remedial options. The revenue/cost imbalance could be attributed to a slumping business cycle or to firm-based weaknesses. Both possibilities are captured in the subsequent Step 3, which turns on the narrower question of whether revenue exceeds operating costs only.

Once again a positive answer points toward short-term continuance, despite the overall loss incurred. However, the failure to meet even operating costs prompts the Step 4 question on the chance for eventual recovery. If severe losses are judged temporary, then Step 5a inquires about the level of

Figure 1-3

Decision Flow Chart for Firm Rationalization

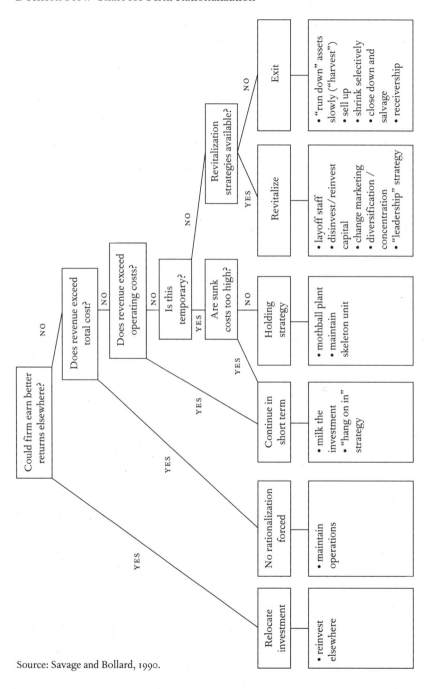

Source: Savage and Bollard, 1990.

sunk costs. High sunk costs (fixed investment) argue for short-term production while conditions clarify; low sunk costs point toward holding, but not operating, the assets. On the other hand, Step 5b, which arises when losses appear indefinite, forces the question of whether revitalization is an option. A positive response triggers a move toward renewal by some combination of classic modernization measures: sharpening the product focus, renewing capital goods, reducing labour, or other. However, the negative verdict opens the way to final closure as losses are cut for good.

In the chapters below, a number of leading firms within our selected industries will be subject to close examination. In order to facilitate comparison over time, *within* industry and *across* industries, a pair of firms will be analyzed in each case. The choices effectively illustrate the striking range of cross-firm variations that affect corporate strategy and outcome. In the pulp and paper sector, the focus is on Domtar, a prominent multi-plant eastern Canadian producer, and Stora Enso, a global leader that has long operated a single integrated facility in the province of Nova Scotia. With origins far from the forest sector, Domtar was transformed into a paper giant through postwar mergers and acquisitions under the sponsorship of a parent holding company. Its subsequent withdrawal paved the way for the Quebec state to achieve controlling interest and to treat Domtar as a strategic provincial asset. The Port Hawkesbury plant was Stora's first investment outside of Europe. In many ways, it represents a classic branch plant operation. However, changing business and political conditions have imposed a continuing set of challenges that has seen the Nova Scotia unit transform itself structurally several times.

For the steel industry, attention centres on Stelco, the leading national integrated producer, and IPSCO, one of the earliest and most successful "mini-mill" operators. Originating in a classic turn-of-century industrial consolidation, Stelco evolved from a high-volume specialist producer into a fully diversified primary producer that has dominated the national market by superior productivity. Only in the past 15 years has Stelco experienced the convulsions of the mature product phase, aggravated by high capital costs, volatile industrial relations, and aggressive international trade strategies. For its part, at the initiative of a socialist prairie government, IPSCO originated as a mixed private capital/state capital venture in regional manufacturing. Utilizing scrap metal inputs in an electric arc furnace, it specialized in high-demand construction and pipeline products that could be delivered competitively in regional markets. In a fascinating contrast, IPSCO has prospered during the turbulent years since the 1982 recession in Canada and has gone on to become a dynamic multinational producer.

Finally in passenger air transport, we compare the two pillars of a national duopoly, Air Canada and Canadian Airlines International (CAI). Today only one survives. Yet for more than a decade, at a time of progressive industry liberalization, the two were locked in a bitter battle for national and international market share. It was not always thus. Air Canada was born in the 1930s as a national state enterprise to ensure a strong domestic presence in a newly emerging industry. Allotted a dominant market share in a tightly regulated sector, the state airline anchored the industry for almost half a century, through periods of virtual monopoly and periods of managed competition with "regional" competitors, before the decisive liberalization began in the mid-1980s. One critical outcome was the privatization of Air Canada. In contrast, CAI's genealogy is the pioneer "bush" airlines, which evolved later into regional carriers, with a strong element of provincial state sponsorship in most regions of the country. Later, with the effective deregulation of the airline sector, CAI pushed forward through mergers and acquisitions to become a national rival to Air Canada. Yet, rather than fulfilling the optimistic expectations bred of US experience, these two firms stumbled to new levels of crisis. They entered the 1991 recession with crippling debt levels, haemorrhaging red ink, and locked in a desperate battle for domestic market share, thus triggering sharp debates about the need for re-regulation. The decade saw several attempts at domestic merger and several bids for alliances and equity partnerships with US carriers before the ultimate denouement of 1999. The airline sector experienced more turbulence in one decade than in the entire preceding half-century.

Exploring the Politics of Industries

In a market economy, firms exist to earn profits by selling goods and services. Regardless of particular lines of production, some features are common to all profit-seeking enterprises in a capitalist system based on the private ownership of property. This is what gives meaning to the notion of common political interests in the "private sector," the "market system," or the "enterprise culture," all terms commonly used by students of the business world.

At the same time, a series of distinctions quickly intrudes into this broad domain of business. There are many key respects in which businesses clearly are not alike. Large firms often have interests that are distinct and even at odds with smaller firms. For example, business giants can plan their investment and production moves and can target designated profit levels with a confidence smaller firms will never know. Similarly, firms that produce and sell on an international scale have concerns that are materially different from firms focussing on a national or regional market. The domestically based firm certainly has its hands full coping with financiers, suppliers, employees, customers, and state authorities in the home market. However, for the international firm, these are multiplied many times, together with the challenges of foreign exchange fluctuations, international capital flow, and wage and productivity differentials. A third range of distinctions is based on industry and even subindustry groups, as we have already begun to discuss.

It is only reasonable, then, to assume that the capitalist business environment is in some respects common to all firms (generating certain interests of business as a whole), in other respects varied by product group (generating interests at the industry level), and in still other respects unique to each firm (reflected in specific corporate interests). While it is important to be sensitive to the interaction of all three levels, a particular challenge of this section is to identify some key *properties* of the industry environment, which by virtue of shared circumstance shape the political interests of business blocs.

One response is to insist that each industry or even subindustry is unique. This implies that the foundations for political mobilization may be as diverse as the number of industries examined. Without question there are certain residual qualities associated with each industry, arising from the particularities of its activity. This point is made routinely by industry spokesmen who both celebrate their special fraternity as "steelmen," "oilmen," or "bankers," while at the same time they lament that politicians, the public, and even other businessmen will never appreciate the exceptional circumstances they face. Certainly such features need to be acknowledged, and we will explore the special character of three such industries below. It does not follow, however, that the prime political characteristics of industry are so idiosyncratic as to defy comparison.

Several bodies of literature suggest the contrary by arguing that the decisive features shaping industry interests revolve around a limited number of variations on core properties. We will now examine three such approaches: industrial classification and market structure, the business-industry dichotomy, and the product cycle.

Industry Classification

One of the difficulties in identifying firms according to productive activity is the complex range of commodities normally generated by a single enterprise. Only in traditional microeconomic theory, it seems, is a firm still expected to produce a single item, such as the tailor's pin, for sale in a homogeneous pin market. In the contemporary economy, familiar commodities such as steel sheet, paper rolls, or airline seats quickly dissolve into a bewildering range of products familiar mainly to the professionals who make, sell, and buy them. Few laypersons can explain the difference between a steel ingot and a bloom or a bar. While most of us could confidently separate newsprint from bonded paper, could we be so sure when confronted with light-coated, non-bleached pulp, and reproduction grades? Finally, though one airplane seat from Toronto to Tel Aviv may appear to be the same as another in all respects save ticket price, this masks a plethora of distinctions involving ticket class, level of in-flight service, scheduled or charter carriers, routing and connections, reservation networks, and frequent-flyer credits.

In order to sort out the scope of industries and to trace the relationships among them, it is common to rely on a standard typology such as the North American Industry Classification (NAIC). This identifies both the series of broad industry groups and the more specific subgroups that coincide most closely with the definition of an industry as a "producer of like goods or services." Since the NAIC is constructed on the basis of *product lines*, it highlights sequences of production in both intermediate and final commodities and services. Moreover, through it, each company can then be located in its industry according to the particular range of commodities it generates, as well as those it does not.

The classification begins by designating a series of broad industry groups, denoted by three-digit numbers. Within each group, more specific product lines are identified by four-, five-, and six-digit numbers. The latter are regarded generally as the most reliable denominators of product competition. Although few firms confine their activities to a single five- or six-digit category, company activities can be mapped against the NAIC grid for a measure of specialization and diversification. Table 1-2 illustrates the main product groups within our three chosen industries.

Political representation can also be organized at the level of the industry or subindustry. Here the most common organizational form is the trade association, whose membership is drawn from the range of firms active in the industry. While some such groups date back to the nineteenth century, the proliferation of trade associations is a more recent development. A brief glance through the "associations" listings in the Ottawa telephone directory reveals how complex a web has been spun. In Canada, Isaiah Litvak identified a domain of some 640 national trade associations, with many more operating at the provincial or local level (Litvak 1982). By far the greatest number were established in the decades since World War II. Most associations operate with a rather small staff, num-

Table 1-2

North American Industry Classification, 1997

NAIC		Details
322	**Paper Manufacturing**	
3221	Pulp, paper, and paperboard mills	
32211	Pulp mills	
322111	Mechanical pulp mills	Mechanical and semi-chemical pulp mfg., not integrated
322112	Chemical pulp mills	Chemical pulp mfg., not integrated with papermaking
32212	Paper mills	
322121	Paper (except newsprint) mills	Integrated pulp and paper, bags, stationery, and sanitary products
322122	Newsprint mills	
32213	Paperboard mills	Building paper and coated paperboard
3222	Corrugated paper product mfg.	
32221	Paperboard container mfg.	
322211	Corrugated and solid fibre box mfg.	
322212	Folding paperboard box mfg.	
322219	Other paperboard container mfg.	
32222	Paper bag and coated and treated paper mfg.	
32223	Stationery product mfg.	
32229	Other converted paper product mfg.	
322291	Sanitary paper product mfg.	
322299	All other converted paper product mfg.	
331	**Primary Metal Manufacturing**	
33111	Iron and steel mills and ferro-alloy mfg.	Steel mills with blast furnaces (including electric arc)
3312	Steel product mfg from purchased steel	
33121	Iron and steel pipes and tubes mfg.	Steel pipe and tube
33122	Rolling and drawing of purchased steel	
331221	Cold-rolled steel shape mfg.	Rolled products made from purchased steel
331222	Steel wire drawing	Steel wire, cable, and rope made in drawing mills, and wire products (nails, staples, barbed wire, etc)
481	**Air Transportation**	
4811	Scheduled air transportation	All except air courier business
4812	Non-scheduled air transportation	All except non-scheduled air courier business
481214	Non-scheduled chartered air transportation	
481215	Non-scheduled specialty flying services	Combination of flying services

Source: Statistics Canada 1997.

bering as few as four to six people. Their functions range from the collection of industry statistics, often of a highly sensitive and confidential nature, and the dissemination of technical and professional data through conferences and publications to the more explicitly political role of monitoring governmental action and lobbying on behalf of the membership. Reflecting this diversity of roles, the senior staff of trade associations tends to be recruited both from the executive ranks of the industry and from the public service.

There is seldom a simple one-to-one correlation between industry category and trade association. More commonly, a three- or even four-digit category may reveal a number of rather specialized associations, representing discrete subindustries (Coleman 1988). Where a single umbrella association successfully spans a broader industry group, this normally involves significant delegation to working subsectors or associational committees so that common positions and concerted leverage can be focussed wherever possible. An effective trade association suggests that horizontal business rivalries have been overcome, for certain purposes, by the force of shared concerns. The latter may stem from vertical business tensions (e.g., steel producers vs. steel importers or steel distributors), from class tensions (e.g., British Columbia forest industry employers vs. their employees), or from shared interests in state policy (e.g., airline operators vs. federal regulatory authorities).

The general dynamics of organizing and maintaining solidarity among association members have been extensively studied. One approach treats this as a function of rational choice, where prospective members weigh the tangible advantages of membership against the associated costs. Mancur Olsen has explored this in terms of the "logic of collective action" (Olson 1971). One familiar dilemma, faced by all voluntary organizations, is the problem of the "free rider." Since many of the gains won by interest groups are collective benefits, which cannot be restricted to members alone, the free-riding person or firm stands to benefit from an organization's efforts without paying any of the costs.

An association can overcome the free-rider problem, at least in part, by dispensing selective benefits of sufficient value that the calculus of membership becomes positive. Not surprisingly, many trade associations originated in efforts to provide services to members in fields such as labour relations (resisting unions) and insurance and freight discounts. In fact, one of the most valuable positions that an association can achieve is a monopoly over the delivery of services for which membership defines eligibility. Other normal concerns of trade associations, such as product standard-setting, collecting and dispensing industry performance data, and negotiating regulatory matters with the state — for instance, in areas such as tariff rates, advertising codes, or product labelling—are cases of more diffuse benefits that may flow to firms whether they belong or not.

The possibility of wider business *coalitions*, constructed on the basis of associational alliances, should also be noted. Certainly these remain the exception rather than the rule, since they depend on a rather exacting set of circumstances, namely, a situation where the interests of several industry sectors, and their organized bodies, are similarly affected by a business practice or state intervention. One intriguing recent case involved a set of industries that shared an interest in reducing their telecommunication costs. Under the banner of the Canadian Business Telecommunications Users Association, they intervened in regulatory hearings to support a move towards a more competitive long-distance telecommunications industry. Solidarities may also develop along vertical lines of business linkage. When the Canadian Steel Producers Association (CSPA) sought access to the highly independent South Carolina Senator Strom Thurmond, whose congressional committee controlled a key trade bill, the introduction was ultimately secured through the Michelin Tire Company. Michelin was both a leading customer for Canadian steel for tire beltings and a leading South Carolina corporate employer (Kymlicka 1987). To this extent, another clue to coalition potential may be found in the input-output tables that quantify the scale and direction of inter-industry linkages (Hirschmann 1958).

David Baron draws an insightful contrast between two forms of political collective action. Horizontal action involves the mobilization of firms within an industry that recognize their common interests, while vertical action is based on the mobilization of diverse interests along the length of the core firm's (or industry's) "rent chain." This includes employees, suppliers, dealers and distributors, customers, lenders, and any others who earn "rents" (or commercial returns) as a consequence of the core firm's actions (Baron 1999). So, for example, the unionized employees of integrated steel companies have developed into staunch allies of those firms in struggles for defensive trade policy support in Washington. Equally, the North American auto manufacturers have collaborated with the steel industry in designs for a new generation, ultra-light, and strong auto body prototype. At the same time, however, North American import brokers, steel distributors, and commodity steel consumers have resisted protectionist measures for big US steel, since the import rent chain offers significant cost advantages to *this* vertical coalition.

Market Structure

The NAIC provides a workable, if imperfect, denominator for the identification of industry groups. Yet in order to appreciate the character of particular industries, each must be addressed not only as a set but also as a configuration of firms. One approach is through a theory of market structures, which yields a set of dynamic properties and hypotheses about industry interests. This approach, central to the industrial organization subdiscipline within economics, stresses the importance of relations *among firms*. Although firms within a particular industry segment are never identically situated, they must all operate within the confines of a market structure whose defining characteristics include the range of competitors, the mechanism for pricing, and the distribution of market shares, all creating a strong element of interdependence among competitors (Green 1990).

The archetypal market structures can be arranged on a continuum, stretching from Adam Smith's classic model of atomistic competition at one end to the pure monopoly at the other. In the former case, firms are price-takers in a setting where the way is open for new entry or exit and where market share is widely dispersed. In the latter, the single supplier can set whatever price the market will bear as the barriers to entry are prohibitive; excess (monopoly) profits are the result. Between these two stands the oligopoly form, a case in which a small number of large firms dominate production under conditions of high entry barriers and minimal price competition.

These market structures can be linked to industry groups through statistical measurement. By calculating the distribution of market shares on the NAIC statistical base, a surrogate measure for firm-concentration can be derived. One frequently used calculation is the concentration ratio, which captures the proportion of total industry (NAIC group) production held by the largest four (CR_4) or eight (CR_8) firms. This calculation is normally done with NAIC data for output or sales, although profits, assets, and employment are also used. As an approximate measure, a CR_4 coefficient exceeding .50 is treated as a loose oligopoly, while a coefficient exceeding .75 suggests a tight oligopoly. Several qualifications must be kept in mind. Since the CR measurement is based on domestic production rather than total supply, it is most accurate when imports are minimal, and it overstates the degree of concentration when imports are significant. The concentration ratio is also silent on the relationship among the largest firms, as a CR_4 of .80 may conceal a single dominant price leader or a more even oligopoly, as the following data suggest.

Consider the data on national steel output in Table 1-3. The CR_4 coefficients are so tightly grouped as to suggest similar industry structures. Yet the breakout by leading firms within national industries shows that this is clearly not the case. Italy is a case of single-firm dominance, France a duopoly, and Canada a rather tight oligopoly. Not only will competitive dynamics differ across the three cases but so, one expects, will the political horizons of the firms.

Table 1-3

Shares of National Output in Crude Steel, 1969

Italy		France		Canada	
Italsider	57.5%	Wendel-Sidelor	35.0%	Stelco	36.2%
Falek	6.9%	Usinor	33.3%	Dofasco	22.4%
Fiat	7.6%	Chiers	3.7%	Algoma	17.0%
		Normandie	3.3%	Sysco	—
CR_4	72.0%	CR_4	75.3%	CR_4	75.6%

Source: Cockerill 1974.

Alfred Eichner has advanced the term "megacorp" to capture the modern type of firm integral to oligopolized sectors (Eichner 1985). In early twentieth-century North America, a series of factors facilitated its emergence. These included new technologies, which enabled supply to surge dramatically beyond demand, thus encouraging market control; changes in corporation law to permit generalized incorporations; the development of sophisticated financial markets to capitalize the giant new companies; and advances in management techniques, especially in accounting practices and cost control. This combination of new capabilities propelled the oligopoly firm into new types of planning through the formulation and implementation of business strategies.

Industry versus Business

In a helpful analysis of comparative business performance, Magaziner and Reich draw a critical distinction between an "industry" and a "business." Most industries will contain several businesses, distinguished according to the markets they serve, the production methods they apply, the cost structures they face, or the barriers they pose to new firm entry. So, for example, the airplane manufacturing industry includes both Boeing and Cessna, although the former is part of the commercial passenger jet business while the latter is part of the small plane business. Magaziner and Reich suggest further that the salient characteristics of businesses do not often coincide with the boundaries of industries as revealed by an industrial classification. Consequently, "when considering the competitive process, one should be wary of superficially discussing this or that industry or subindustry" (Magaziner and Reich 1982, 68).

Since the competitive position of most businesses hinges on one or two crucial characteristics, these should be kept in mind. One broad category consists of sheltered businesses, which are insulated from import and export considerations by virtue of a domestic or local market focus. In many cases, such businesses can survive and even prosper despite serious productivity failings, since no alternative exists. Other service industries, such as finance, telecommunications, and transportation, are centres of high-value activity, and their growing internationalization has pushed the question of free trade in commercial services to the top of the WTO agenda.

On the other side stands a number of classes of traded businesses, where international competition is a central factor and competitive advantage must be sought. One set of traded businesses is most sensitive to *raw material* costs. In pulp production, for example, the costs of wood fibre can account for as much as 40 per cent of total input costs, while in primary steel the combined impact of iron ore and coal is also substantial. In another set of traded businesses, competitive position turns on keeping *labour* costs low. This has been a traditional feature of the textile and clothing businesses, though, as we shall see, it also figures in many of the core heavy industries today. A third set of businesses is distinguished by *complex-factor* costs. Their production process is sufficiently complex, involving closely articulated stages and continual adjustment and change, that the key lies in the coordination of expert systems. This is the case when a modern steel plant utilizes continuous casting methods to supply a number of finishing mills, or when an air carrier network must coordinate with precision a complex set of scheduled commitments each day.

Not only do such characteristics help explain the shared concerns of firms within a business, but they also reflect directly on the business strategies adopted by individual firms. More generally, business characteristics also carry crucial implications for the framing and delivery of effective state policy. Any superficial assumption that firms conform to a single model is highly dubious. The simple nostrums that access to capital or low interest rates or tax reductions will automatically improve the performance of any enterprise must be rejected and replaced by a more discriminating assessment of the fundamentals of discrete businesses:

> There are no direct relationships between aggregate levels of investment and competitive success in the economy as a whole, or in broad industries such as steel, automobiles or semi-conductors. Higher levels of investment can be useful to attain better productivity only if correct strategies are followed in order to gain a competitive advantage. Otherwise there is a double loss—more funds have been diverted from other uses, and yet competitive deterioration and the accompanying economic stagnation continue. (Magaziner and Reich 1982, 106)

At the same time, the competitive foundations of "business" sketched above take in vast corporate terrains. It is certainly possible that such features may be translated into shared sets of political and policy preferences. One such scheme has been advanced by Thomas Ferguson as a theory of "industrial partisan preferences" (Ferguson 1984). In the same way that citizen voters make their electoral choice, Ferguson suggests that large firms and major industries are drawn into party coalitions. But unlike voters, industry choice is structured first and foremost by business calculations. Two variables are particularly powerful predictors: the labour sensitivity of the industry, measured by the ratio of wages to total value-added, and the trade sensitivity of the industry, measured by the strength or weakness of domestic firms facing international competition.

This can be mapped by plotting labour sensitivity along one axis and trade sensitivity along the other. By its coefficients, each firm and industry can be located in historical time in one of the four resulting quadrants. Ferguson contends that, for much of the twentieth century, two clusters or coalitions predominated, one based in domestic-oriented, labour-intensive industries and the other in internationally traded, capital-intensive industries. Out of this emerged the modern US party system, whose fortunes have echoed the shifting magnitudes of the two business blocs. The turn of 1900 saw the Republicans emerge dominant. President William McKinley and his corporate mentor, Mark Hanna, forged the Republican business coalition known as the "system of '96." Rooted in protective tariff and anti-union politics, this bloc remained hegemonic well into the 1920s. Then, in little more than a decade, the order of dominance was reversed. Franklin Roosevelt successfully articulated the basis of the "new deal" investment coalition, which drew capital-intensive and international trading industries behind the Democrats in an alliance that continued until the 1970s. In later work, Ferguson and Rogers extended this model to include another historic business coalition. Ronald Reagan's neo-conservative Republicanism was erected on a set of new industries including real estate, resource extraction, microelectronics, and finance (Ferguson and Rogers 1981, 1986). Its policy coordinates revived the "system of '96" anti-labour perspectives, though its trade stance remained liberal and internationalist.

Product Cycle

For a generation, the model of the product cycle has shaped business thinking about industrial change. Although it can be linked to the paradigm of market structure, the concept of product cycle is rooted in a different set of variables.

It was originated to explain investment and trade patterns in terms of the changing role of innovation, scale economies, and knowledge/uncertainty in the decision-making process. This was first advanced by Raymond Vernon in an effort to move beyond the neo-classical explanation of economic behaviour in terms of comparative costs (Vernon 1966). Vernon sought to address the Leontief paradox, by which the capital-to-labour ratio in US export industries was lower than the ratio for domestic industries.

Vernon's alternative explanation posited a stage-like model for product development. It stressed the singular circumstances of a newly developed product whose specifications and production techniques had yet to be standardized (i.e., conditions of high uncertainty and high expenditure on product development). The intensive and costly product development research tended to be concentrated in the advanced capitalist metropoles, particularly in the US. By contrast, once a degree of standardization had been achieved within the industry, the next stage saw the rapid growth in demand and supply for the successful product (Vernon, 1971). The competitive focus shifted to the refining of cost advantages in production, as a number of differentiated product lines were explored. In addition, the sites of production began to spread into other advanced capitalist economies. In yet a further stage, the production of basic commodities becomes sufficiently standardized that many of the earlier barriers dissolve and the product can be manufactured on a wide scale in locations around the globe. Vernon suggested that export strength was greatest in the later products, in contrast to those at earlier, more capital intensive, phases of the cycle. Finally, a fourth stage can be identified in which demand for a product declines due to obsolescence or displacement (Levitt 1965).

Overall then, a product, as distinct from an industry or business, follows a development path that varies across its life-cycle. Many refinements and variations have been offered over the years, with empirical applications to specific industries, contrasts between extended and compacted product cycles, and the possibility of cycles being renewed or reset by the impact of new technologies (Wells 1972). Questions have been raised about the continuing relevance of this theory in an age of globalized production and finance; however, it does offer a useful perspective on the dynamic qualities of products, which affect all firms within the product market.

In an intriguing extension of the Vernon model, James Kurth has explored its political ramifications in determining industry coalitions and state policy responses (Kurth 1979). Assuming that most, if not all, rival firms move together along the product cycle curve, they may acquire a set of shared interests according to phase. This suggests that, at a given moment, a capitalist economy must seethe with the ferment of contradictory industry interests, rooted in dozens of product cycles out of phase with one another. Even more important, it helps to account for fundamental changes in the policy outlook of a single industry over time, as its economic needs evolve. Yesterday's free-trading sector can predictably emerge as tomorrow's defensive protectionist lobby as the product cycle unfolds.

Kurth compares the evolving political agendas of the textile, steel, and automobile industries in many of the leading capitalist economies. Not only does he document the changing policy agendas as each industry follows its product cycle, he also illustrates the historical shifts in dominant position over time. Finally, he observes that, while single industries are unlikely again to dominate as completely as textiles, steel, and autos did in the past, policy conflicts and coalitions among major industries will be a continuing feature of economic policy-making.

Building on Kurth, it is possible to hypothesize a general set of correlations between product cycle phase and public policy response. One version is set out in Table 1-4.

Exploring the Politics of the State

The sections above demonstrate that politics is a natural and inevitable part of both enterprise and industry activity. We turn now to the role of state authorities in shaping the micropolitical environment. This raises questions about how the modern state is organized, how it relates to business (and non-business) interests, and where we should expect to find micropolitical relationships.

Just as a firm or an industry can be broken down into its constituent interests, so too can the state. We refer to the state most generally as the source of sovereign authority, distinguished by its ability to take and enforce decisions of universal application backed by the force of law. Depending upon constitutional provisions, this may reside with a single national authority (a unitary state), or it may be divided between national and regional authorities (a federal state). Either way, a state needs to be seen as a complex structure, consisting of many distinct and specialized components (Miliband 1973). These may be defined by their functional specialities, such as providing over-all leadership (executive), law-making authority (legislature), administrative expertise (bureaucracy), law adjudication (courts), or civil order (police and army). Each of these branches or subsystems contributes to the functioning of the overall state system, and each follows its own rules and procedures.

Table 1-4

Product Cycle Phases and Policy Correlates

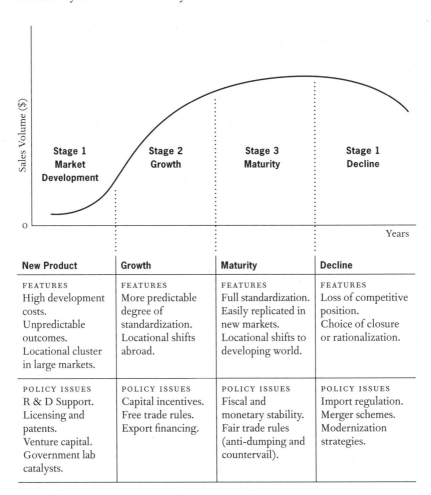

New Product	Growth	Maturity	Decline
FEATURES High development costs. Unpredictable outcomes. Locational cluster in large markets.	FEATURES More predictable degree of standardization. Locational shifts abroad.	FEATURES Full standardization. Easily replicated in new markets. Locational shifts to developing world.	FEATURES Loss of competitive position. Choice of closure or rationalization.
POLICY ISSUES R & D Support. Licensing and patents. Venture capital. Government lab catalysts.	POLICY ISSUES Capital incentives. Free trade rules. Export financing.	POLICY ISSUES Fiscal and monetary stability. Fair trade rules (anti-dumping and countervail).	POLICY ISSUES Import regulation. Merger schemes. Modernization strategies.

Another way to portray the state is in terms of its substantive operations, which reflect the wide variety of policy areas and programs undertaken at any point in time. There are several useful indicators of a state's overall commitments. For example, the collected laws or statutes provide the legal authority for a state to intervene. New laws are constantly being formulated and existing laws amended or rescinded. Indeed, the ongoing body of constitutional and statutory provisions forms both a framework for political activity and a target for it. Second, the organizational structure of the bureaucracy indicates how these responsibilities are packaged and delivered. Departments and agencies are the key agents of policy formulation and delivery, and it is through senior staff and elected officials that accountability is maintained. Departments tend to exhibit their own complexity through the many specialized branches and divisions that contribute to their overall mandates. Thus, problems of political coordination challenge the state both at peak levels and in each constituent unit.

The phenomenon of deliberate and coordinated state action in pursuit of policy goals is often captured in the concept of state capacity. One early formulation sought to distinguish between strong and weak states according to their abilities in planning and coordinating responses to societal pressures. Over time, however, the notion of a generalized state capacity came into question. Distinctions were drawn according to policy goals, such as accelerating economic growth, mobilizing political consent, securing social order, implementing foreign policy, or managing industrial adjustment. Each of these required distinct policy instruments and political coalitions to proceed. As Weiss notes, "it may well be that it is the *unevenness* of state capacity that is most significant for understanding state behaviour in certain spheres" (Weiss 1998, 4). Our interest lies principally in the efforts of firms and industries to organize politically and to engage state authorities on issues of growth and adjustment.

To raise the question of overall state structures is to underline the qualities of political coherence and stability. Ordered hierarchical structures are capable of disciplined and coordinated activity. But for what purposes, and for whose interests, is this pursued? Not surprisingly, business has played a prominent role in shaping state policies since colonial times.

It is common to distinguish a set of historical eras of state intervention according to the predominant political strategies of each. For example, McBride suggests a series of national policies in Canada since 1867, based on the overarching roles of the state in supporting the shifting interests of organized capital. The first of these, covering the half-century to the 1930s and the Depression, sponsored an integrated industrial economy behind a finely balanced tariff structure. The second phase, unfolding after World War II, combined the Keynesian strategy of fiscal management for full employment with a collaborative federal-provincial welfare state, all within a progressively liberalized trade and investment environment. A third strategy, asserted in the 1970s, aimed to remedy the accumulated sectoral imbalances through an integrated industrial strategy linked to both resource and manufacturing sectors. It failed, however, at least in part due to the narrowness of the domestic business beneficiary segment (McBride, 2001).

Most significant, for our purposes, is the role of organized business interests in determining the political prospects of each economic project.

The complex character of Canadian business virtually precludes unanimity on overarching economic questions. However, for each successful strategy supportive interests were mobilized by at least one peak business association, whose intervention proved critical to the outcomes. The embryonic Canadian Manufacturers Association (CMA) was a key ally of the Macdonald Conservatives in the formative stages of the first national policy. The CMA played a crucial role in reconciling inter-industry differences as the manufacturing tariff was formulated and adjusted. For the second era, which rose from the desperate economic conditions of the 1930s and was refined during the centrally managed wartime economy, business opinion was once again divided. Sectors such as banking and manufacturing proved relatively receptive while resources, importing, and exporting were resistant (Finkel 1979). However, the Montreal Board of Trade seemed to play a particularly instrumental role before the Rowell-Sirois Commission when possible new parameters of state intervention were advanced.

The third and, in this case, abortive national policy arose in the contradictory economic conditions of the 1970s. This strategy centred on a series of post-Keynesian programs combining regulation of foreign equity investment, wage and price controls to contain escalating inflation, and a national energy policy. Significantly, this phase was the least strongly supported in the business sector. Despite the efforts of businessman-politician Walter Gordon and his Institute for Economic Policy, and support from the Canadian-owned fraction of the petroleum industry, the post-Keynesian program was broadly construed as threatening to capital.

Indeed a crucial new organization emerged in 1976 in the form of the Business Council on National Issues (BCNI). Drawing its membership from the chief executives of the top 150 Canadian corporations, it was a diversified representation of big business, both domestic- and foreign-owned. Not only did the BCNI (renamed the Canadian Council of Chief Executives, the CCCE, in 2002) resist state expansionism, it advanced a counter-strategy of comprehensive market-based liberalization. Over the course of the next decade, it galvanized business opinion behind a continentalist trade and investment regime.

The current neo-liberal phase of macroeconomic strategy emerged in the mid-1980s, although the Mulroney Conservatives came to power in 1984 without explicitly embracing this program. However, by the next year, the galvanizing impact of peak big business in organizing the wider Canadian capitalist bloc (Langille 1987) registered in two respects. First, it formed the policy backbone of the Macdonald Royal Commission Report on Canada's Economic Prospects, which itself provided a roadmap for neo-liberal reform.

Second, it provided crucial political endorsements for the Mulroney government's free trade negotiating initiative with Washington.

Clearly, the politics of peak business associations operates on a separate level from industry and firm-based politics. The former tends to focus upon macropolitics and conditioning framework policies in such fields as fiscal and monetary policy, competition policy, and trade. However the potential relevance of peak business politics should not be overlooked in the chapters to follow. It is a fact that firms from paper, steel, and airlines were all members of the BCNI in the formative years and remain active in the CCCE today (refer to Table 1-1).

It is virtually impossible, for the reasons sketched above, for a firm or industry interest to engage with a state as a whole. It is more common for both business and wider societal interests to target more specialized state authorities at the level of department, branch, program, or agency where ongoing relationships can be built around shared concerns for the design, approval, and delivery of policy. In this sense, each state program, decision, and material and symbolic allocation is a contestable target. The convergence of diverse and conflicting interests in this way makes such state operations inherently political.

Policy Analysis

In each of the chapters below, we will be analyzing significant public policy conflicts that go to the heart of industry politics. But first it is necessary to establish some tools to identify, dissect, and compare policy measures. Perhaps the most basic unit of policy is the *decision*, a formally executed choice that sets action in motion. Any government concludes hundreds of decisions each day on differing scales of significance. For example, we will explore a 1971 decision by the government of Canada to regulate the content of pulp and paper effluent discharge, a 1987 decision that directed Canadian National Railway to purchase steel rails from Sydney Steel, and a 1999 decision to permit Air Canada's proposed takeover of CAI.

However, every formal decision is the culmination of a more complicated *process* by which an issue emerges, finds definition, is subject to conflicting claims and pressures, and is eventually resolved. This too is evident in the three examples above. It becomes clear that the effluent decision was rooted in several prior policy streams, one involving the long-standing administration of the federal Fisheries Act and another based on more recent concerns with the health effects of chemical discharge. For this sort of regulatory issue, scientific input plays a major role. In the case of steel rails, we see the politics of procurement preferences played out against commercial pressures in a sharply competitive market. Here a relatively modest expenditure decision provoked spirited interventions from two provincial governments as well as from the US, evolving in the process into a divisive regional and continental issue. Finally, the Air Canada takeover marked the culmination of more than a decade of tortuous political manoeuvring that followed domestic deregulation. In each case, a complex mix of offices and agencies sought to shape the final decisions.

It should also be noted that few policy decisions stand strictly on their own. More typically, a decision is part of a wider cluster of related measures which can be described as a *policy area*. A combination of factors, including bureaucratic process, legal authority, professional norms, and client group connections, contribute to consistency and continuity within an area. Additional complications will arise if an issue or decision is located at the intersection of several policy areas. For example, pulp mill emissions form part of the regulatory area of industrial pollution. Yet pulp and paper's role as a leading export and manufacturing industry brings considerations from the industrial policy area into play in determining pollution policy results. How then should we conceptualize the role of state actors in such disputes?

Policy Cycle

A useful perspective on state activity can be found in the policy cycle model, which acknowledges the sequential nature of decision-making, while at the same time it highlights the characteristic features associated with successive stages (Howlett and Ramesh 1995; Pal 1992). Many versions of the policy cycle have been advanced over the years, but they share several features in common. First, policy issues or decisions tend to develop in a broadly predictable fashion. Second, the policy process can be distinguished by a series of functional steps or stages, each possessing its own political dynamic. The value of such a framework is to allow any issue to be located on a template that can highlight key analytic questions. It also offers some tactical insights on the primary challenges associated with each stage. The key steps are set out in Figure 1-4.

The policy cycle begins with the process of *issue definition*. This tends to coincide with the time when a political problem begins to gain visibility and attract attention in the world of government. It is important to appreciate that no political issue springs to life spontaneously. It must be given form, and this is normally the product of organized effort. Issue definition involves not only

Figure 1-4

Model of the Policy Cycle

Issue Definition	Agenda Setting	Policy Formulation	Policy Adoption	Policy Implementation
i) Issue Recognition ii) Delimitation iii) Issue salience	i) Ordering priorities ii) Actionable short lists iii) Review and Re-ordering	i) From Ends to Means ii) Technical design iii) Choice of policy instruments	i) Formal Approval	i) Putting into Effect ii) Discretionary Applications iii) Negotiated Outcomes
Signifying Events				
Framing the Meaning	Strategic Prime Ministership	Bureaucratic Politics		The Complexity of Joint Action
Reinterpretation	Issue Fatigue	Executive Federalism		Strategic Litigation
Issue Triangulation	Crowding Out			
LEAD ACTORS Ministers Interest Groups Think Tanks Public Inquiries Mass Media Trade Associations	LEAD ACTORS PM/Cabinet Central Agencies Ministerial Staff	LEAD ACTORS Departments Regulatory Agencies Consultants	LEAD ACTORS Cabinet/ Ministers Central Agencies Parliament	LEAD ACTORS Departments Regulatory Agencies Judiciary
	SOURCES Throne Speech Election Platform Budget Speech Crisis Management		SOURCES Cabinet Decision Ministerial Order Regulatory Decision Statutory Enactment International Treaty	

the identification of a problem, but also the delineation of boundaries and the specification of what, exactly, is problematic. A policy "topic" does not equal an issue. Public opinion polls report frequently that "the economy" is a leading "issue" for citizens of Canada. But closer reflection shows this does not qualify as a defined issue in the technical sense. The latter requires a specification of what precisely within the vast field of "the economy" needs attention. Is it a matter of unemployment being too high, of capital investment too low, or of taxes skewed unacceptably in one direction or another? Trade unions would argue the first, business managers the second, and tax-paying citizens the third. Not all can be accommodated at the same time, and the battle for issue definition is one for discursive primacy.

Clearly, issue definition can be a hotly contested stage. In part it involves a war of ideas, albeit a war directed less at discovering objective truths than in capturing the attention, and framing the terms, of public understanding. Not surprisingly, this affords a prime role to the "idea merchants" based in public inquiries, research institutes, think tanks, and media outlets (Dobuzinskis 1996; Micklethwaite and Wooldridge 1996). On their own, however, ideas are seldom sufficient. They normally offer an interpretation of the problem together with a prescription of solutions. Influential ideas are also connected, in an opportunistic fashion, with expressive events — such as currency crises, corporate bankruptcies, hostile takeovers, and workplace strikes — that can signify, embody, or confirm a certain definition. Equally, the shape or substance of issues can be transformed through tactical persuasion. Bill Clinton's political advisor, Dick Morris, practiced the technique of "triangulation." This holds that the path to a desired issue definition is not always the straight line from "A" (current) to "B" (desired). Often, in fact, the successful transition runs through an intermediate position "C," which sets the stage for the ultimate attainment of desired position "B" (Morris 1997).

Public life is always swamped with contending issues. Few, however, are chosen. The intervening filter is aptly captured by the *agenda-setting* stage of the policy cycle. It is widely believed, in high political circles, that a cabinet minister can effectively manage no more than four or five prominent issues over the term of a government (Osbaldeston 1988). To obtain a successful outcome, each issue will require sustained attention and repeated focus. When Pierre Trudeau returned to power in 1980, he was advised to pursue a "strategic prime ministership" that concentrated on a few key priorities (Clarkson and McCall 1994, 158). Trudeau chose the constitution, energy and industrial policy, and foreign policy review. The same advice applies for senior bureaucratic managers. Given such limited room at the top, the challenge for issue advocates is to crack this privileged lineup. While the competition is always stiff, there are a plethora of short-lists to work with. Consider, for example, the multiple agendas corresponding to the national, provincial, territorial, and municipal governments. Add to this the variable agendas prepared within departments, boards, and agencies of each level of government. For example, the steel industry agenda may be dominated by export trade disputes, the air transport agenda by problems of domestic monopoly dominance, and the pulp and paper agenda by excess capacity and soft prices.

The process of agenda-setting is a complicated one and should not be seen simply in terms of interest group advocacy (Kingdon 1984). Any change in the shape or order of an agenda requires that past priorities be displaced, and the champions of the status quo can be expected to marshal all of their available resources to resist. It is also important to make allowance for the politically unexpected. A good deal of policy involves crisis management, and crises have a way of commanding attention independent of political preference. Just as agendas can be deliberately ordered, they can be reordered as a result of changing circumstance or shifting balances of political forces. Perhaps the best way to approach agenda-setting is to view it as an amalgam of elements distilled in an unsteady laboratory.

The third stage of the policy cycle involves *policy formulation*. By this point, both the conceptual frame and the need for action have likely been agreed upon. Attention shifts to the development of a means to achieve the designated goals. Not surprisingly, at this point the initiative also shifts to the bureaucratic subsystem for the selection of policy instruments or mechanisms of delivery. It is interesting to note the change in central actors that can occur between the opening pair of stages and those that follow. Interest groups and think tanks can be instrumental in the preliminary events, but the actual distillation of instruments is dominated by technocratic officials. Agreement in principle is fine, the saying goes, but the devil is in the detail. Consequently, this often provides an important opening for bureaucratic politics and the unequal structures of representation that this allows.

Stage four involves *policy adoption*, or the formal approval of a course of action. Typically, this involves a cabinet decision, a ministerial order, a statutory enactment, or a regulatory decision. The focus on legal instruments is a distinguishing feature of this stage. Cabinet may authorize new initiatives under existing legislation, ministers and regulators may exercise authority delegated by prior statute, or Parliament may be involved for the making of new laws. While the executive and legislative elites are central to this stage, it is striking to see how late this occurs in the policy process, and how many critical conflicts and choices have already been taken by this point.

Once formal approval has been secured, the initiative shifts to *policy implementation*. Once again the civil service comes to the fore, since program delivery is one of its most significant functions. It should never be assumed, however, that political tensions have been resolved fully by the time this point is reached. Indeed, skilled practitioners of the lobbying arts know that the policy implementation stage marks the opening of a second front. Much can happen between decision and delivery: delay is a political resource in its own right. Budget constraints may slow the pace of implementation, and discretion may be delegated to officials for technical work and fine tuning. All such adjustments can alter both the sense and the impact of a previously agreed-upon program. In addition, it is only when implementation begins that the judicial arena normally becomes operative. Strategic litigation offers the prospective "losers" yet another opportunity to delay or derail the intended policy. Interests that are affected adversely by a public policy measure can only launch a court challenge after it is a *fait accompli*.

Pressman and Wildavsky argue that the politics of policy implementation is burdened by "the complexity of joint action." They put it this way: "we are initially surprised because we do not begin to appreciate the number of steps involved, the number of participants whose preferences have to be taken into account, the number of separate decisions that are part of what we think of as a single one. Least of all do we appreciate the geometric growth of interdependencies over time ..." (Pressman and Wildavsky 1973, 93). Later in this book we will encounter a number of graphic illustrations. The enactment of pollution effluent standards for pulp and paper mills began in 1971 and was significantly revised in the early 1990s. For both firms and state agencies, it was the experience of applying the first-generation rules that played a major role in shaping second-generation strategies. In the Canadian steel sector, a key reason for support of the 1989 free trade deal was its anticipated defence against exaggerated and punitive trade disruptions on the US side. Following ratification, however, Canadian steelmakers were dismayed to find that there was very little change in the machinations of the US trade bureaucracy. Finally, the policy decision to deregulate the domestic Canadian air market was adopted in 1984. However, the implementation of that decision has occupied the entire period since, generating a host of complications which themselves have altered the nature of the deregulatory plan. In each case, the unanticipated and unmanageable policy complications arising with implementation served to transform the political alignments that had initially underwritten these decisions.

Despite the many virtues of a policy cycle model, there are definite limitations that must be kept in mind. It is crucial to remember that such frameworks are indicative rather than definitive. Not every issue will progress in an orderly sequence through the designated stages. Not every policy initiative will survive through to the final phases. Some will be blocked or abandoned or postponed along the way, so that the sequence may be arrested or reset at a future date. Nonetheless, an accurate assessment of the state of play for any policy is an essential place to start. Here the cycle model provides a useful point of reference.

Policy Networks

When organized interest groups were first recognized as prime agents of political influence, there was a tendency by analysts to view state authorities as empty vessels that could be shaped and filled according to group pressure. Political leaders were portrayed as passive or reactive figures—gauging the sum of the pressure vectors, treating them as a democratic consensus, and registering the net result in formal political decisions. In this view, the groups or coalitions of groups that brought the most resources to bear would prevail over their less effective rivals.

Over time, however, the premise that the state, functioning as an umpire or a weather vane, stood outside or above politics was abandoned as both naive and false. Instead, the state came to be recognized as an integral part of the political process. State agents were seen to possess organizational or institutional interests of their own, which figured independently in the calculus of policy choice. For example, Graham Allison popularized the notion of "governmental" or "bureaucratic" politics. Every branch and agency was a potential repository of interests, which stemmed in part from program mandate or purpose, in part from shared, predominating technical and professional outlooks and in part from the necessity of defending program mandates against rival state agencies. Allison coined the famous slogan that "where you stand depends on where you sit" (Allison 1971), meaning that your position (where you stand) on an issue depends principally upon your office location (where you sit) in the state hierarchy.

The recognition of potentially complex configurations of political interests, housed within the state, marked a major analytic step forward. Yet Allison cautioned that pure forms of bureaucratic politics, in which intra-state interests dominated policy outcomes, were relatively rare and were most common in fields such as foreign and defence policy, where high levels of internal secrecy served to insulate the policy process. It stood to reason that the more typical domestic policy-making context would see societal interests closely interlaced with bureaucratic dynamics. Furthermore, the multiple channels of access presented by the modern administrative state underlined the potential advantage of mobilizing at the sectoral, industry, or even enterprise level of interest.

Both marxists and neo-pluralists advanced versions of a state-society paradigm. Heavily influenced by Nicos Poulantzas, many marxists spoke of the "relative autonomy" of state structures. Rather than being absolutely controlled by economically dominant classes, the state functioned with a degree of freedom to fashion political compromises among conflicting class forces (Poulantzas 1975). A national or regional state was conceptualized as a hierarchy of departments and agencies into which a variety of class interests could be incorporated. The state thus served as an unequal structure of representation, which allowed the adjustment of political allocations according to shifts in the balance of social class relations through time (Mahon 1977).

The neo-pluralists devoted greater attention to the "process" dimensions of group-state interaction. Many advanced a concept of the "policy network," a set of relations that linked organized interests to one or more state institutions in an issue area. This model has been applied, with considerable success, to the study of business politics (Atkinson and Coleman 1989). Network analysis suggests that permutations of groups and states fall into a number of typological forms, distinguished by certain conditioning attributes. For interest groups, the *level of mobilization* of the aggregate constituency drew particular attention. What pattern of group organization, density of group membership, and ability to bind members to policy agreements prevails? For the state, a key index is the *level of autonomy* possessed by administrative structures. Are they sharply focussed around clear mandates and legislated authority, do they enjoy support from the political executive, and are they buoyed by a professional ethos? The point here is to gauge the capacity of officials to act independently of group pressures. A second index is the *level of concentration* of state structures. How many levels of jurisdiction and how many bureaux or agencies are involved in a policy area? In short, this asks about the prospects for coherent and coordinated state action. An illustrative typology of policy networks is presented in Table 1-5.

Understood in this way, networks form a significant part of the policy context for micropolitics. Just as market structure and product cycle constrain the path of action without tightly determining specific outcomes, so too does

Table 1-5

Typology of Policy Networks

Mobilization of Business Interests	State Structure			
	High autonomy/ High concentration	Low autonomy/ High concentration	High autonomy/ Low concentration	Low autonomy/ Low concentration
Low	state-directed	pressure pluralism	pressure pluralism	parentela pluralism
High	concertation	clientele pluralism	corporatism	industry-dominant pressure pluralism

State-directed: the state defines policy strategies with major implications for business, without consultation.

Pressure pluralism: business interests are fragmented and rivalrous, firms and associations engage in policy advocacy rather than power-sharing.

Clientele pluralism: the state delegates initiatives to business associations, which pursue goals sanctioned by state policy.

Parentela pluralism: close relationships between owners/managers of firms and the dominant political party.

Concertation: business interests (industry or peak levels) share power with state authorities.

Corporatism: producer groups (business and labour interests) share power with state authorities.

Source: Atkinson and Coleman 1989.

policy network. Business interest groups form only part of the network domain, albeit a rich and intriguing part of the overall picture. Their functions as business advocates, the variations in associational patterns, and the underlying structures of industry and subindustry activity have prompted extended study of their policy dynamics. All three of the feature industries in this study are thick with associational interests. Particular attention will be paid to identifying the leading associational groups and their contributions in representing shared sectoral interests. It should also be remembered that state agents are necessary partners in policy networks. It is important to understand how the state presence is configured in particular sectors. Does it involve a single department or branch within a department or a broader constellation of rival bureaucracies? How many distinct policy networks are operative within a given policy sector? What sort of hierarchical relations bind the constituents together and what sort of pressures force them apart?

For example, the paper industry has been distinguished by a high degree of sector mobilization. The one dominant trade association, the Canadian Pulp and Paper Association (CPPA), was formed during the first wave of highly capitalized expansion in 1913 and is characterized by a high membership density. Under this umbrella, sections were formed to address technical and policy concerns about woodland harvesting, mill technology, labour relations, and markets. This functional decentralization went so far as to provide dedicated journals and section conferences for the various concerns. With a staff of more than 80, the CPPA became one of the largest and most sophisticated trade associations in the nation. On the other hand, fissures within the state resulted in a pattern of low concentration. Federal departments, including industry, trade, finance, fisheries, and environment, have in the past and continue to play roles in pulp and paper policy, as do provincial forest resource and environment agencies. Low autonomy is another dimension. This points to a major potential dilemma, as the state seemingly lacks the capacities to assist a sector experiencing structural stress.

In steel, by contrast, there was no tradition of collective mobilization of producers at the associational level. It was not until the 1980s that the CSPA was formed, in the midst of profound steel trade tensions with the US. The leading steel firms could avail themselves of association channels through the CMA or Canadian Chamber of Commerce; however, the more common pattern was firm-specific representation, as in the Tariff Board inquiry of the 1950s or the steel profits inquiry of the 1970s. The pattern of state organization was once again diffuse. Today, important mandates fall to the federal departments of industry, revenue, finance, and environment, and the Canadian International Trade Tribunal (CITT). Not surprisingly, the steel sector operates in a pressured pluralist environment, with little tradition of clientele networking.

The passenger air sector has been densely mobilized since the formation of the Canadian Air Transport Association (CATA) in the 1930s. This group played a strong role during the regulated phase of the business (1937-84). There were also, however, forces inclined toward firm-specific advocacy, since each company required separate regulatory approval for its routes, fares, and aircraft. Since domestic deregulation, the shakeout of major carriers has deprived the CATA of a major membership layer. In any event, there was little room for a common industry voice on the Air Canada/CAI battles of the 1990s. However, the CATA remains a strong voice on the technical side of the industry and as an umbrella for more than 100 smaller commercial operators. It is important to appreciate that networks are dynamic. Not only can they evolve as sector conditions change and policy coordinates shift, they may vary by subindustry as well. For example, domestic air transport evolved from a clientelistic, regulated network to a pluralistic, open-entry regime over the 1980-90 period. By contrast, international air transport remains locked in a state-driven pattern of bilateral air service agreements.

In air transport, there have been dramatic turns in the national state presence since 1980. During the classic age of domestic air regulation, the Ministry of Transport was the one-stop centre for policy delivery, and the minister's office was a powerful clearing house. It is true that operational responsibility was dispersed internally, between the department, which handled infrastructural functions such as air traffic control, airport management, and air safety, and the Canadian Transport Commission, which licensed carriers, routes, and equipment. Compare it, however, with the present arrangement in which air traffic control has been privatized and independent local airport authorities provide terminal services while an autonomous air safety board handles safety issues, and a scaled-down Canadian Transportation Agency follows the "fit, willing, and able" approach to licensing.

It is hardly surprising to find business interest groups at the centre of industry policy networks. It would be wrong, however, to stop a network analysis at this stage. To reiterate a point already made, trade associations are only part of a policy network. It is important to acknowledge three other potential levels of involvement. One concerns firms operating as discrete political interests rather than through a collective agent. This applies by definition to companies that decline membership in trade groups. It also extends to firms that, for whatever reason, find the need to break with the common voice in order to assert overriding particular concerns. As we will see in the chapters below, the leading firms in oligopolistic industries are eminently capable of mounting enterprise-based government relations campaigns. A second level involves organized non-business interests. Some obvious candidates here are trade unions and consumers. The former have had a decisive impact on the timing and extent of restructuring efforts in all three industries. The latter have precipitated change through their ability to translate shifting consumer preferences for cheaper products and new product attributes to both industry and state. Perhaps the most important emerging layer of non-business interests involves the non-governmental organizations, particularly the environmental advocates (ENGOs), that have emerged over the last generation. This is particularly evident in pulp and paper, where effluent disposal has become a continuing political frontier.

It is worth noting that struggles over particular issues are usually rooted, to some degree, in prior political structures. Thus, they are best understood within the context of ongoing network relations and policy frameworks inherited from prior decisions. Almost invariably, new issues are advanced and received against the template of past battles and settlements. Forces seeking the review, revision, or replacement of established policy commitments must acknowledge the existence of entrenched interests and interpret the map of alliances and cleavages through which change becomes possible.

Conclusion

This chapter has advanced several frameworks intended to clarify the dynamics of micropolitics. The challenge is to put them to work on particular firms and industries. Keep in mind that the political agendas of business are sweeping and dynamic. The short lists of politically sensitive issues are not always very brief, and the "hot-button" concerns are notoriously transitory. At best, the chapters below provide points of entry and perspective on these changing domains. However, the historical background, the structural portraits, corporate case studies, and public policy profiles will hopefully offer readers some pathways into wider and deeper applications.

Key Terms and Relationships

Board of Directors: Corporate officers elected by shareholders to oversee company business and policy as executed by senior executives. Works through committees and as a plenary body.

Capacity (state): The ability of a state (or state agency) to mobilize, co-ordinate, or regulate investment effort and industrial change in concert with economic interests.

Corporate Governance: Institutional arrangements and procedures by which an organization takes formal decisions and maintains accountability.

Decision: The basic unit of policy-making both in firms and states; a formal choice of action by an organization. Decisions occur at successive stages of the policy cycle.

Employees: Persons hired by private or public organizations to work at assigned tasks, either for salary (monthly or annual rate) or wage (hourly or weekly rate).

Horizontal and vertical political coalitions: Alliances created among interests within a single industry and among interests along a rent chain, respectively.

Institutional shareholders: Those investment firms, including banks, financial dealers, mutual funds, pension funds, that actively buy, hold, and sell shares in large volumes so as to achieve maximum short-term profit. A key audience for corporate executives.

Junk bonds: High-interest bonds issued by firms that are unable to attract more conventional "investment grade" buyers. Pioneered in the 1980s by dealers such as Drexel Burnham Lambert.

Liquidation value: Value of a firm's assets if sold, in whole or part, to interested buyers after business closure.

Money markets: The financial markets for currencies and trading in foreign exchange.

Network (policy): A configuration of organized interests and state agents that are linked together in a specific policy field.

Oligopolies, Oligopolistic industries: Cases where a limited number of large firms hold a disproportionate share of output or sales and can exert concerted market power over business conditions.

Rent chain: A set of linked business transactions between firms exchanging inputs and outputs, in which margins or profits (rent) are extracted at each stage.

Share-to-asset ratio: A business value measure that compares the market value of a firm's shares (or equity invested) to the market value of its assets (or property).

CHAPTER TWO

Pulp and Paper Politics

The Industry at the Millennium

For the pulp and paper industry, the twentieth century closed with an extended wave of corporate mergers and acquisitions. In Scandinavia, in North America, and on a trans-Atlantic basis, leading firms looked to takeovers as the favoured growth strategy. Canada was at the forefront of this process.

The billion-dollar deals began in 1996 when Montreal-based Donahue bought control of Quno for $1.4b. The following year, Canadian and world newsprint giant Abitibi purchased Stone Consolidated of Chicago ($4.4b), while Domtar took over E.B. Eddy ($1b). In 1998, the US Bowater Corp. purchased Avenor (formerly CP Forest Industries) for $2.4b, becoming North America's second largest newsprint producer. A year later another US firm, Weyerhaeuser Corp., absorbed Canada's largest forest products company, Vancouver-based MacMillan Bloedel ($2.4b). The year 1999 also saw Smurfit-Stone (US) purchase St. Laurent Paperboard of Montreal, while Fletcher Challenge Canada bought the Pacific-rim pulp and paper assets of its parent, Fletcher Paper. The new millennium was only months old when Abitibi paid $5.8b for control of Donahue. Norway's Norske Skog followed with a takeover of Fletcher Challenge Canada and, with the later addition of Pacifica Papers, became one of the leading paper firms in the country. In the spring of 2001, Domtar spent $1.6b to acquire a cluster of US-based uncoated paper mills from Georgia-Pacific.

Such transactions were part of a wider world consolidation wave in pulp and paper. The Scandinavian giants Stora (Sweden) and Enso-Guzeit (Finland) merged in 1998 to form Stora Enso, the global number two enterprise. That same year, world leader International Paper took over US rival Union Camp ($6.3b), while Jefferson-Smurfit did the same with Stone

Container. Globally, the year 2000 opened with Finland's UPM Kymmene fighting off International Paper for the prize of Champion Paper. In the tissue sector, Georgia-Pacific Corp.'s $11b takeover of Fort James dwarfed all prior deals.

In some respects these may seem to be surprising developments. The 1990s were not kind to pulp and paper. Commodity prices were sluggish, and corporate profits were slim, at a time when capitalism at large experienced one of its longest cyclical booms since World War II. It was often remarked, ruefully by forest-sector interests and dismissively by others, that long-term government bonds offered better financial returns than did pulp and paper stocks. When the leading equity markets took off into record growth in 1995, the pulp and paper share indexes remained slow to stagnant.

This points to a deeper set of problems. In the northern hemisphere at least, leading segments such as market pulp and newsprint seem to have entered the mature phase of their product cycles. Aggregate demand has, for the most part, levelled off. Global production of market pulp, however, has exploded with the opening of world-leading scale facilities in Latin America and Asia. While fluctuations will always be part of the demand for pulp, newsprint, coated paper, tissue, and board, and while these segments will sometimes move in contradictory fashions, the major structural problems are found on the supply side.

Pulp and paper is the most capital intensive of all manufacturing industries, and new plants add productive capacity in massive quantums of 250,000 to 500,000 tonnes per year. Almost all industry segments experienced bursts of new capacity in the 1990s. Equally important, these were not matched by corresponding closures of older, high-cost facilities. The result was oversupply, soft markets, and thin profits. Pricing patterns are captured in Figure 2-1 below. One of the most effective ways to respond to this, in a market setting, is by corporate consolidation through mergers and acquisitions and subsequent rationalization by closing marginal and non-core facilities. Perhaps the puzzle is not why the takeover wave happened, but why it took so long to arrive.

At the same time, it should not be assumed that the pulp and paper industry was a uniformly static or passive environment. The 1980s and 1990s saw many of its core business practices called into question. Process technologies were subject to political challenge, with accelerated public concern for the health threats posed by plant effluent. First-generation pollution abatement techniques seemed ill-equipped as the number of toxic substances mushroomed, and the search for solutions shifted from end-of-pipe to closed-system design. Another dramatic development involved the rise of the recycling movement, which targeted paper

Figure 2-1

Pulp and Paper Prices, 1992-2004

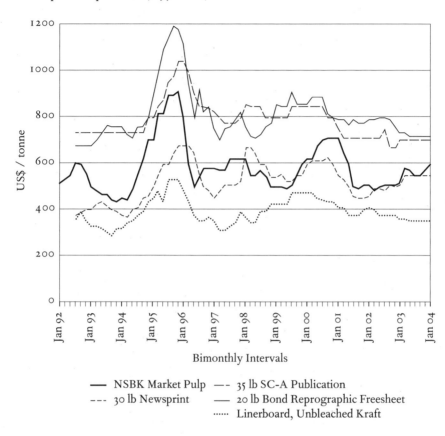

Compiled from *Pulp and Paper Week*, 14 July 1997; 19 January 1998; 11 March 2002; 26 January 2004; 9 February 2004. All prices f.o.b. eastern United States.

products as a prime landfill problem while calling for industrial reuse. In only a few years, firms were forced to change their pulping and papermaking practices to accommodate a variety of pressures from consumers and state regulators. A third generic challenge was directed not at the manufacturing but the timber harvesting stage. Opponents of industrial clear-cutting orchestrated successful campaigns aimed at ending this practice on grounds of ecological disruption and non-sustainability. At a time when greater economies were being sought at the woodlands stage, firms looked for protection to third-party certification of sustainable forest management systems.

Overall, this combination of cyclical maturity and segment, or firm-based, volatility drives the micropolitics of Canadian pulp and paper in the new millennium. In the sections below, we examine these forces in greater detail.

Overview

Perhaps because so many of its manufacturing facilities are tucked away in the forested hinterlands of the nation, pulp and paper is not often recognized as one of Canada's largest manufacturing sectors. At the peak of its last industry cycle, in the late 1990s, this sector embraced more than 180 pulp, paper, and paperboard establishments (mills), located in all provinces save Prince Edward Island. The value of its output was almost C$32b in 1999, 6.5 per cent of all manufacturing. Of this, almost C$23b took the form of exports, accounting for 9 per cent of total exports by value. And of course the returns from sales abroad contributed the largest positive industrial share to the nation's balance of international payments. Finally, some 103,000 workers were employed in the NAIC's 322 group of pulp and paper operations in the largest full-time wage force of all manufacturing industries (Canada 2002d).

The relative obscurity of this industry may be due, in part, to the mundane appearance of its products. Certainly they lack the physical drama of steel-making or jet air transport. Few people outside of the plants would recognize pulp in either the wet or dry varieties if they saw it. For its part, paper might appear to be a rather simple, low-value commodity, useful mainly for print-ing newspapers, which are also an inexpensive, throw-away product, or for wrapping groceries. Moreover, the rise of plastic packaging might suggest that the uses of paper are growing fewer with time. In fact, world demand for pulp and paper products continues to grow, doubling in the past 20 years, and to assume a variety of new forms. As we will see, one of the challenges for Canadian producers over the past generation has been to adjust production capability and marketing strategy to capitalize on newer, higher-value prod-ucts at a time when the more "standard commodity" production is shifting from Scandinavia and Canada to more southerly locations.

In fact, the industrial processes of separating cellulose fibre from wood and fashioning it into new products are scientifically complex and extremely demanding. The resulting range of pulp and paper products is indeed vast. As an intermediate product, pulp can be derived from a variety of different wood inputs: hardwood and softwood, boreal forest or tropical forest, or combi-nations of these. It can also be made by different processes. Groundwood yields the highest volume of fibres per unit while chemical treatments yield the purest, longest fibres. The resulting product may retain its natural, often brownish, colour, or it may be bleached white, depending on the ultimate destination of the pulp. Once again, pulps vary according to the bleaching agents that are used, from traditional chemicals such as chlorine to the newer oxygen and peroxide processes.

The worldwide demand for cellulose fibre is so vast as to support an entire industry of free-standing pulp mills (Carrere and Lohmann 1996). Here the final product is "market pulp," sold on the open market to papermaking firms which choose to purchase rather than produce their fibre inputs. Today the largest pulp operations in the tropics have capacities of 1 million tonnes per year or more. In Canada, a full 34 per cent of total industry output consists of marketable wood pulps (Canada 1997b). There are even spot trading markets for commodity pulp futures and derivatives to enable suppliers and buyers to hedge the uncertainty of forward delivery prices (Lindeberg 1996; Kenny 1997).

The alternative manufacturing strategy involves some degree of vertical integration within the firm to link the pulping and the papermaking activities. Here again, paper is anything but a simple commodity. The most familiar form of paper may still be newsprint. With the rise of literacy and mass communication in the late nineteenth century, the demand for a cheap, light paper that could hold ink led to a massive investment in both Canada and Scandinavia. Canada remains the world's largest producer of newsprint, a status that confers both benefits and liabilities, as we shall see. Not surprisingly, newsprint accounts for the second largest share—32 per cent in 1996—of pulp and paper output in Canada.

Other paper products account for the remaining 34 per cent of Canadian output. These include book, magazine, copy, and writing paper (the "printing and writing" papers); tissue and sanitary paper; packaging and paperboard; and building papers. Since each of these includes a plethora of specific products, many firms concentrate on a single sector. For example, Kimberly-Clark (US) is the world's largest tissue and sanitary paper producer, while Abitibi-Stone (Canada) is the world's largest newsprint manufacturer. Domtar Inc., whose strategy is analyzed in some detail below, has pursued a more balanced product mix until very recently. Our other featured firm, Stora Forest Industries of Sweden, entered Canada as a stand-alone pulp mill and later diversified, in stages, first to newsprint and then to calendered printing papers.

It is also important to underline the extent to which the pulp and paper industry is regionally differentiated within Canada. This is captured on Map 2-1. Although the forest *sector* is the largest element in the British Columbia

Map 2-1

Location of Pulp and Paper Mills in Canada, 1988

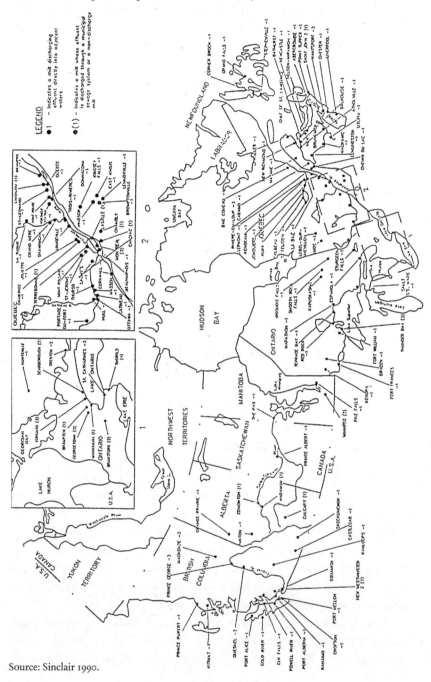

Source: Sinclair 1990.

economy, the pulp and paper *industry* runs a definite second to lumbering, and the pulp fibre supply chain is linked directly to sawmill outputs. It is in eastern Canada that pulp and paper predominates, particularly in the provinces of Quebec, Ontario, and New Brunswick. Over the past 15 years, Alberta has seen an explosion of new investment, putting it into fifth place nationally (Pratt and Urquhart 1994). The other prairie and Atlantic provinces have a far more modest share of the national market, though pulp and paper often figures as an extremely powerful force within them (Clancy 2001). Such variations have many roots. In part, they reflect the quality and extent of forest inputs, the positional advantages in specific markets, and even the timing of provincial entry into the industry. But, as an independent variable, regionalization inevitably affects the character of industry politics, since the level of shared interests, as well as the degree of solidarity and rivalry among firms, fluctuate accordingly. In an industry where both provincial and national state authorities play central roles, this is a recipe for varied and intricate political relations.

One final characteristic is worthy of note at the outset. Few Canadian industries have found themselves so intensely affected by the rise of contemporary environmental movements, an experience for which most firms were poorly prepared and slow to respond. Bolstered by traditional clientelistic relations to state agencies, as well as a complacency linked perhaps to its strategic prominence in the nation's economic life, pulp and paper behaved in a reactive as distinct from a pro-active way until very recently. A host of environmental policy issues have been raised over the past 25 years, questioning many of the central operational practices of the industry. These range from the effect of pesticide and herbicide chemical sprays on human health and forest ecology and the effect of clear-cut logging on forest renewal and landscape integrity to the damage resulting from mill effluent discharge into rivers, lakes, oceans, and air, and the sustainability of one-time product use and disposal (May 1998). These questions have meant mounting levels of state regulation and public criticism at a time when business conditions were already extremely volatile.

After discussing the historical evolution of pulp and paper in Canada, this chapter examines the structural basis of the industry today, with particular attention to strategic positioning. Two firms are then analyzed in detail. Domtar Inc. is a diversified eastern Canadian firm which plays a particularly strong role in the province of Quebec. Stora Forest Industries (SFI) is an industry leader in Nova Scotia. The former is a Canadian-owned enterprise that has changed hands several times over the past generation. The latter is a Scandinavian-owned subsidiary of a global forest products giant, whose story speaks to the complexity of life within a multinational firm. Both Domtar and SFI provide strong illustrations of the force of internal corporate politics. Finally, two case studies of public policy in pulp and paper are featured. One involves the federal pulp and paper modernization program of the 1970s and 1980s, which was designed to upgrade process technologies and machinery at older mill sites, particularly in eastern Canada, at a time of heightened international competition. The other probes the design and application of pollution prevention regulations by federal and provincial authorities as one of the most serious political challenges to any primary resource sector in Canada.

Historical Development

The art of papermaking is as old as civilization and has relied, in turn, on a wide variety of materials. The root of the word "paper" is the ancient Egyptian "papyrus," a reed which could be cut, layered, soaked, and pounded into sheets from which scroll parchments were fashioned. The Chinese were using paper by 105 AD, and from there it spread west to Asia Minor. Until the Moors brought Chinese techniques to Europe after 1100 AD, the only other common form of parchment was vellum, made from calf- or lambskin. In fifteenth-century Europe, after Gutenberg invented the printing press, most paper was based on rag fibre, whose strength and durability remains unsurpassed today. However, the mounting demand for writing and printing paper periodically outran the supply, and Europe experienced repeated "paper famines."

It was only in the nineteenth century that wood became the predominant fibre source in the industrial capitalist economies of Europe and North America, while other vegetable cellulose remained important in parts of the colonial and postcolonial world. In addition to trees, cellulose could be drawn from such natural products as flax, cotton, jute, and sisal, as well as from agricultural waste products taken from fields in such forms as straw, grass, corn stalks, bamboo, or sugar cane. Mineral elements such as asbestos have also furnished fibre on occasion, though it has restricted uses and appeal today for obvious reasons of health. In the contemporary period, waste materials have once again emerged as a prime source, with recycled paper and cardboard proving to be increasingly important.

One obstacle to the exploitation of wood fibre lay in the difficulty of extracting cellulose from solid logs. While wood cellulose can make up 50 per cent of a log, it is embedded by a resinous adhesive known as lignin, which amounts to another 30 per cent of log volume, with the balance accounted by extractable aromatic oils and carbohydrates (Greenpeace International 1990).

During the nineteenth century, a number of techniques were pioneered in Europe to separate fibre from lignin. The first involved "mechanical" operations. Initially, the cellulose fibres were torn out of their densely compacted mass by a literal "grinding to a pulp" in heavy stone mills. However, this punishing process required huge amounts of energy, which were initially supplied by water-power and later by hydroelectricity. Today, groundwood techniques remain important in pulp production, not the least because this process yields a high proportion of cellulose fibre relative to solid log volume. At the same time, since grinding breaks the fibres and lignin extraction is not complete, groundwood pulp products have weaker fibre networks and little "tear strength," as well as a tendency to yellow after prolonged exposure to light. Consequently, mechanical fibre is most commonly used in newsprint grade products. In recent decades, mechanical pulping has undergone another revolution by combining high temperature heating with mechanical separation (now done with metal disks) in a process known as thermo-mechanical pulping or TMP. A further hybrid technology is chemi-thermo-mechanical pulping or CTMP.

Also in the nineteenth century, advances in chemical engineering opened alternative routes. The first of these was the sulphite treatment, in which wood chips were cooked in sulphuric acid. Up until World War II, the sulphite process was predominant among the chemical techniques. However, in the period since, it has been increasingly displaced by sulphate, or kraft (the German word for "strong"), methods. In this case, the chips are cooked in caustic soda by a system that recovers more than 95 per cent of the chemical for reuse. Since chemical pulps eliminate virtually all of the lignin, the resulting pulp consists of whole fibres that are strong and resistant to yellowing.

Chemical mills are most easily identifiable from their sulphur dioxide emissions, known as "rotten egg odour" in resource towns across the continent. While chemical processes are far less energy intensive, they produce huge volumes of complex waste compounds, which must either be recovered for future use or disposed off-site. Over the past several decades, this industrial effluent has raised a thicket of environmental impact questions. By one estimate, the pulp and paper industry accounts for a full 9 per cent of total industrial pollutants in Canada (Sinclair 1990). Chemical pulping techniques have generated an additional wrinkle in the form of "bleaching." Chemical pulps emerge from the cooking processes in various shades of brown colour and, while this does not matter in the manufacture of some paper products, it is unacceptable for others. Consequently, many chemical pulps must be bleached white through the use of an additional set of chemicals. Until recently, the most popular bleaching agent was chlorine, until it was implicated in a variety of serious illnesses and listed as a toxic substance.

The interplay between pulping techniques and raw wood supply can also be complex. Just as all manufactured fibres are not equal, neither are all species of trees. Deciduous or broad-leaved trees are primarily hardwoods. In North America, the pulp-oriented hardwoods include birch, beech, and aspen, while in more tropical climates the favoured hardwoods are eucalyptus and acacia. From the point of view of silviculture (i.e., afforestation) there is a dramatic contrast between the two types. While the growing period for northern hardwoods from seed or seedling to harvestable age may be as high as 60 to 80 years, the tropical hardwoods can reach commercial maturity in as little as 12 to 15 years. Since this permits a far more rapid crop rotation, it is a major factor in the recent global shift toward tropical pulp production. Coniferous, or softwood, species represent the other major type. These are generally needle-based trees such as spruce, fir, and pine. Once again, there are intriguing contrasts among the softwoods, as the northern species such as spruce and fir have rotation periods of 40 to 60 years while various southern pine species, such as the loblolly (US) and radiata (New Zealand), may be half as long.

For commercial purposes, one of the key differences turns on fibre length, as hardwoods tend to be shorter (1-2 mm) while softwoods are longer (3-5 mm). Offsetting this to some degree is the higher fibre content, relative to lignin and oils, in hardwood as opposed to softwood. While many paper products use a combination of hardwood and softwood fibre, the basic fibre stock still tilts toward the latter, capitalizing on its strength properties, particularly when manufactured by chemical techniques. This explains the rapid growth of kraft pulping in Canada, the US, and Scandinavia during the post-World War II era. Today the grade of "northern softwood bleached kraft" (NSBK) remains a benchmark for market pulp products.

All of these factors should be considered as strategic parameters for firms entering and operating in the pulp and paper industry. The choice of site location for a mill is closely connected to the availability and security of wood supply. The choice of manufacturing processes, the linkages between plants and range of final products, is similarly driven. At the same time, firms must also consider industry conditions of present and future demand, of supply and return, which may alter the profile of a mill complex over its 30- to 70-year life.

While many of the scientific and technical breakthroughs for wood pulp production occurred in continental Europe, the industry was drawn to North America and Scandinavia by the apparently unlimited supplies of raw wood found there. Equally, the advent of pulp and paper served to transform the economic significance of the forests that dominated the boreal regions. Whereas in the past the leading commercial species had been pine for sawlogs, attention turned to the smaller scale stock such as spruce and fir. Although rag paper had been manufactured in Quebec since 1805, the first groundwood pulp mill was opened by A.J. Buntin at Valleyfield in 1869. It was soon joined by the first chemical pulp mill, established by Angus and Logan at Windsor Mills, Quebec, in 1864. These "start-ups" marked the beginning of the industrial era of pulp and papermaking in Canada. Nelles contends that F.H. Clergue built the prototype of the modern integrated pulp and paper complex in 1895 at Sault Ste. Marie, Ontario, situating both the groundwood mill and the paper mill near its electrical generating facility and the timber limits (Nelles 1974). Until then, pulp was manufactured at a woods site and then shipped over considerable distances to paper mills located near urban markets.

State policy also had a critical impact on both the pace and the sites of pulp and paper investment. In response to early lobbying from the US paper industry, the US government levied a sizeable duty on paper imports but a low-to-zero duty on pulp. Consequently, some paper companies built plants at border locations on US soil, which offered ready access to either logs or pulp flows from Canada. Fraser Company located its pulp complex in Edmundston, New Brunswick, while its paper mills sat across the Saint John River in Madawaska, Maine. A connecting pipeline transported the wet pulp (the intermediate product) from one side to the other. While such tariff rules determined many of the sites for first-generation paper mills, the situation changed decisively in 1913, when a counter-lobby from the American Newspaper Publishers managed to have the newsprint duty removed. With free trade in newsprint, paper plants could also move abroad to integrated sites in the forest hinterland. By 1918, Canada had become one of the world's leading newsprint exporters (Guthrie 1941). Following the war, the industry experienced its greatest boom ever, as dozens of mills were established in northwestern Ontario, the Ottawa Valley, the Lac St. Jean and St. Maurice regions of Quebec, and New Brunswick. This influence can be seen on Map 2-1. At the same time, the Canadian paper industry became tilted disproportionately toward the production of newsprint-grade papers for export to the US.

Given its proprietary role as the owner of crown forests, the provincial state was also a central force in industrial development. By contrast to the US, state or crown forest ownership is the norm in Canada. Only in Nova Scotia (72 per cent), New Brunswick (50 per cent), and Quebec (12 per cent) is a significant share of the forest owned privately. Consequently, provincial authorities could wield their licensing authority over crown forest lands to shape the pace and direction of forest industry development, as well as the distribution of benefits among the logging, sawmill, and pulp and paper sectors. Initially, many provinces treated their crown forest reserves principally as a source of revenue, by levying license and stumpage (rental use) fees (Gillis and Roach 1986). While it was hoped that manufacturing linkages would follow from the assignment of timber limits, provinces were also willing to tolerate raw log exports. Yet by 1900 a vigorous campaign was being waged across Canada, calling on provincial authorities to insert a "manufacturing condition" on all major timber leases to prevent the export of raw wood. Ontario imposed just such a condition in 1900, as did Quebec in 1905. However, the federal system left considerable room for variation in this respect, since the balance of political forces was never identical in each jurisdiction. In Nova Scotia, for example, the pulpwood embargo was successfully resisted by a coalition of forest landowners and export merchants, and raw wood continued to be exported in significant volumes throughout these critical decades (Parenteau and Sandberg 1995).

Another key dimension of state policy, which had a direct bearing on the pulp and paper sector, given its substantial energy needs, was the provision of hydroelectric power. At stake were both the terms of electrical supply and the ownership of the generating enterprises. The early 1900s saw the beginnings of the "public power" movement, which called for state-owned utilities to assume control of this important new industry. This triggered an epic political struggle in the province of Ontario, pitting some of the largest and most influential blocs of national and international capital against provincial- and municipal-based manufacturing enterprise (Nelles 1974). Ultimately, the public power movement prevailed, and the Ontario Hydro-Electric Power Commission was established in 1910. For different reasons, state-owned power monopolies also emerged in the prairie provinces. Yet electricity remained a private enterprise for decades or even generations in such jurisdictions as Quebec, British Columbia, and the Maritime provinces.

Even where state enterprise emerged dominant, there remained the question of how to service the substantial energy needs of pulp and paper enterprise in remote northern locations. Since mills tended to be built on major waterways to facilitate log collection, the means for locally generated electricity were ready to hand. Yet, with provincial authorities controlling riparian rights to inland waters, paper firms required licenses for water use. As a result, it was common in the negotiation of new projects for hydroelectric rights to be put on the bargaining table along with timber-cutting rights. This explains the common corporate name terms such as "pulp and power" and "power and paper" companies.

Right up to the eve of the Depression, US and Canadian investment consortia continued to plan and open new paper mills in eastern Canada. However the financial crash on Wall Street in October 1929 spelled the abrupt end of boom times. By the early 1930s, firms suffered from declining demand across the range of paper products. Companies such as the Mersey Paper Company in Nova Scotia, which had only opened for business (with spectacular mis-timing) in December 1929, faced huge problems in servicing their debts (Raddall 1979). Those that survived the decade managed to cancel dividends, reschedule debts, and struggle along on severely reduced capacity, while many others disappeared entirely. Some of the survivors were even able to exploit merger and takeover prospects with failing firms, thus extending their timber bases and acquiring modern plant facilities for the day when business recovered. This recalls Joseph Schumpeter's contention that the swings of the business cycle brought "gales of creative destruction" to modern economies (Schumpeter 1939).

While the Depression reached its depth in 1933 and various business had begun to recover by 1936, the pulp and paper industry did not experience another major growth wave until after World War II. Up until that time, the industry remained centred in eastern Canada, with Quebec, Ontario, and New Brunswick as the leading jurisdictions. By contrast, the British Columbia forest sector was still principally a lumber economy, based on the extraordinary Pacific forests. It was only in the 1950s that pulp and paper made dramatic inroads there in response to two developments. One was the shift of forest investment from the British Columbia coastal region toward the province's interior and the north. The other was the growing commercial exploitation of sawmill wastes, as distinct from the pulp log staple in the east, as the prime source of cellulose fibre (Marchak 1982).

For pulp and paper, as for many industries, the close of the 1960s marked the end of a stable and profitable postwar era. Of course the new stagflationary climate of faltering growth and mounting inflation was challenging in itself, but two additional factors complicated life in pulp and paper. First was the rise of world energy prices, rooted in the first OPEC oil crisis of 1972-73, which threatened the entire cost structure of the industry. The second was Washington's decision to float the US dollar in 1974, ending a generation of fixed international exchange arrangements that had underpinned internationally traded businesses. This introduced a major new operating variable for pulp and paper exporters, since abrupt shifts in rates of exchange could wipe out productivity advantages, disrupt traditional supply patterns, and wreak havoc with profit and loss forecasts.

It was also during the 1970s that Canadian state authorities became increasingly concerned with the structure and performance of this front-line manufacturing and export industry. This emerged in a policy context that raised increasingly serious questions about the impact of foreign investment (relatively low for this sector) and about international competitiveness (extremely high for this sector). By 1978 Ottawa had completed an unprecedented business-government consultation toward a new industrial policy, which included the forest sector (Canada 1978a). One outcome was the pulp and paper modernization program, which is explored below in greater detail.

Structure

The aggregate ranking of major firms, according to scale of operations, captures a dimension of power based on asset size or revenue flow. As was mentioned at the beginning of this chapter, Canadian pulp and paper firms have followed the international trend by consolidating into ever larger units. Table 2-1 indicates the size of the ten largest companies in 2002 and compares them to the global leaders. Perhaps surprisingly, given the comparative advantage which flows from superior resource endowments, the largest Canadian firm stood only eighteenth in the world. This is particularly intriguing since Sweden and Finland, with a similar boreal forest base but far less timber volume, place leading firms in the global top ten.

Several factors may help to explain this outcome. The periods of classic concentration for Canadian forest capital were 1918-30 and 1945-65. However, there were really two geographically distinct tracks for merger and takeover activity: one headquartered in eastern Canada and the other on the west coast. Rarely did these fields of operation overlap to create genuinely "national" paper firms. One of the few exceptions was the Argus Corporation, which acquired forest companies in both the east (Domtar) and the west (BC Forest Products). Yet even here, Argus chose to hold these assets in separate portfolios, which were later sold to separate owners.

A further wave of takeovers occurred in the 1970s, but these were qualitatively different, involving the absorption of pulp and paper firms into diversified conglomerates and holding companies. In addition to the Argus case, Canadian Pacific Industries acquired Great Lakes Paper; the Power Corporation (Paul Desmarais) purchased Consolidated Bathurst; Olympia and York (Paul and Albert Reichmann) took control of Abitibi-Price; and Brascan (Charles and Edgar Bronfman) took over first Noranda and then MacMillan Bloedel. One of the few paper-based entrepreneurial

Table 2-1

Leading Pulp and Paper Firms by Sales: World and Canada, 2002

Rank	Pulp, Paper & Conversion Sales (US$)	Company (HQ)	Market Pulp 1000 tonnes	Paper/Board 1000 tonnes
1	$ 20,386.0	International Paper (US)	2,290	13,712 e
2	$ 16,214.0	Georgia-Pacific (US)	1,397	10,586
3	$ 11,880.0	Procter & Gamble (US)	0	1,600 e
4	$ 10,886.2	Stora Enso (Fin/Swe)	900	13,743 e
5	$ 9,057.2	Svenska Cellulose (Swe)	na	9,490
6	$ 8,611.6	Kimberly-Clark (US)	na	4,000 e
7	$ 8,516.3	Oji Paper (Jap)	80	8,170
8	$ 8,433.6	UPM-Kymmene (Fin)	na	10,046
9	$ 7,950.4	Nippon Unipac (Jap)	286	7,910
10	$ 7,910.0	Weyerhaeuser (US)	2,281	8,212
11	$ 7,483.0	Smurfit-Stone (US)	514	7,095
12	$ 6,861.0	MeadWestvaco (US)	0	5,800 e
13	$ 6,186.8	M-real (Fin)	260	5,180
14	$ 4,637.4	Mondi International (SoAfr)	260	4,087
15	$ 4,438.9	Jefferson Smurfit (Ire)	116	3,931
18	$ 3,202.1	Domtar (Can)	716	3,027
20	$ 2,837.8	Abitibi-Consolidated (Can)	106	6,394
27	$ 2,178.1	Cascades (Can)	117	3,058
39	$ 1,300.2	Tembec (Can)	na	1,142
52	$ 944.0	NorskeCanada (Can)	374	1,727
57	$ 774.2	Norampac (Can)	0	1,313
67	$ 600.0	Nexfor (Can)	162	627
86	$ 408.6	Canfor (Can)	na	109
101	$ 308.7	West Fraser Timber (Can)	10,189	na

e is an estimate   na is not available from company

Source: Compiled from *Pulp and Paper International*, 45 (9) (September): 25-34.

giants to emerge during this period was Repap Enterprises. Its architect was George Petty, who, after guiding a cast-off Temiscaming mill into the employee-owned Tembec Inc., went on to assemble a network of mills in New Brunswick, Manitoba, and British Columbia that flourished in the 1980s before tipping into bankruptcy a decade later.

Much research remains to be done to explain the causes and consequences of conglomeration for the Canadian pulp and paper industry, but it seems indisputable that the holding companies were more interested in the earning power of their subsidiary assets than in their long-term growth as free-standing forest-sector giants. This may well have bent their strategic visions in directions not shared by paper-sector managers on their own. Niosi has argued that large-scale Canadian capital came of age during the 1970-85 period, under the supportive policy umbrella of the Canadian state (Niosi 1985). If so, the truncation of the Canadian paper firms may be an integral consequence. Just as the Foreign Investment Review Agency may have facilitated the concentration of Canadian corporate assets in domestic hands, it may also have cut pulp and paper firms off from sources of international capital at a critical expansionary juncture. Of course this began to change after 1984, when the Mulroney government declared that Canada was again "open for business." Soon New Zealand, Japanese, Scandinavian, and US capital made major inroads into the Canadian industry. But just as big Canadian capital turned outward in the era of free trade, the prolonged paper sector downturn of 1990-94 led to the abandonment of many conglomerate holdings. For Canadian firms, a crucial window of expansion had been missed, and the industry leaders slipped from the second to the fourth rank of global operations. Belatedly, this has begun to change, with the late 1990s wave of mergers and takeovers noted at the beginning of this chapter.

While aggregate analysis certainly conveys important aspects of large-scale enterprise, it is not the end of the story. Another perspective comes from examining the market structures of the industry (Canada 2002a). To judge strictly from the aggregate forest industry data, the four-firm concentration ratio (CR_4) for Canada in 1992 was only 25 per cent, suggesting a highly competitive market structure. The corresponding CR_8 figure was 40 per cent. This points to the deficiencies of whole industry data. Firms are not in business to produce and sell "pulp and paper," but to sell a wide variety of specific commercial items, all widely separated in technique and purpose. Consequently, the subindustry offers a more realistic denominator for measuring industry structure. When we disaggregate the data to the level of major product groups, a different picture emerges, as will be seen in the discussion, below, of the market pulp, newsprint, printing and writing, and tissue segments. These more closely approximate what Magaziner and Reich (1982) label as separate "businesses."

The commodity flows documented in Figure 2-2 offer a useful guide to both the connections and the variations. It begins with the principal sources

Figure 2-2

Product Flows from Wood Fibre to Final Paper Products, Canada, 1996

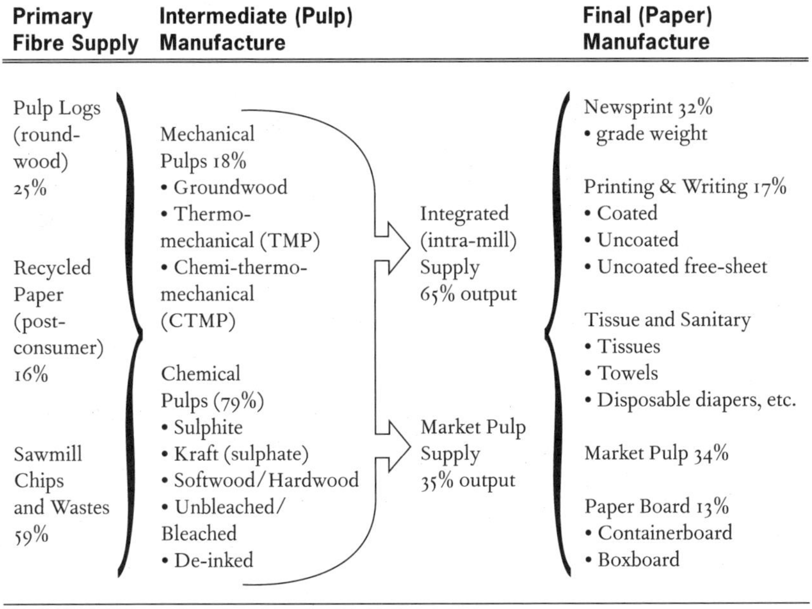

Primary Fibre Supply	Intermediate (Pulp) Manufacture		Final (Paper) Manufacture
Pulp Logs (round-wood) 25%	Mechanical Pulps 18% • Groundwood • Thermo-mechanical (TMP) • Chemi-thermo-mechanical (CTMP)	Integrated (intra-mill) Supply 65% output	Newsprint 32% • grade weight Printing & Writing 17% • Coated • Uncoated • Uncoated free-sheet
Recycled Paper (post-consumer) 16%	Chemical Pulps (79%) • Sulphite • Kraft (sulphate) • Softwood/Hardwood • Unbleached/ Bleached • De-inked	Market Pulp Supply 35% output	Tissue and Sanitary • Tissues • Towels • Disposable diapers, etc. Market Pulp 34% Paper Board 13% • Containerboard • Boxboard
Sawmill Chips and Wastes 59%			

Source: Adapted from *Industry Canada* 2002a.

of primary wood fibre, including the harvesting of sawlogs (large-diameter trees) and pulpwood (smaller tree, generally under 12 inches in diameter). It also involves the production of wood chips, either by mechanical breakdown at the stump in the forest or from the reprocessing of sawmill waste materials such as edgings (sections that are trimmed away as logs are squared). Sawmill by-products have emerged as the leading source of pulp fibre inputs, rising from 10 per cent of wood supply in the 1960s to more than 50 per cent today. These operations are generally the preserve of the woods departments of pulp and paper firms, which plan logging operations on both company lands and crown leases and purchase wood from private suppliers where necessary. While it is not a major focus of this chapter, it is important to realize that the purchase of private timberlands, the acquisition of crown leases, and the procurement of privately grown fibre have all been fields of extended political conflict in Canada (Sandberg 1992). Finally, the rise of recycled fibre needs to be noted. There are three principal sources of post-consumer paper: corrugated board (as much as 50 per cent), office papers (25 per cent), and recycled newspapers (20 per cent). While de-inked and recycled pulp lacks the strength of virgin fibre, it can be quite effective as part of a mixed feedstock, and it is heavily mandated, in both Europe and North America, by law and regulation.

Next we come to the wood pulp and paper group. This begins with the intermediate stage, which transforms solid wood into a variety of pulp products, based on type of fibre (softwood or hardwood) and manufacturing process (mechanical or chemical). While softwood pulps are still preponderant—the NSBK grade is the key global reference price—technical advances have opened a wider range of applications for hardwood pulps and made possible the recent forest boom in northern Alberta.

MARKET PULP

Approximately two-thirds of this pulp remains "captive" inside the mill gate and is transferred in wet form to papermaking machines. However, the final third is sold as the commodity known as "market pulp." As such, it represents one of Canada's areas of export strength, with major markets in the US, western Europe, and Asia. Almost half of the pulp capacity in British Columbia is dedicated to market pulp, yielding a variety of grades of dried and pressed pulp that is shipped to independent paper mills (Canada 1997b). Though Canada accounted for 30 per cent of global market pulp production in 1996, this segment was in considerable flux. Since fibre procurement accounts for half of production costs, market pulp is extremely sensitive to conditions of wood supply. In the 1980s and 1990s, market pulp plants of increasing scale were opened in Latin America and Asia at sites close to tropical groundwood supplies. This expanded capacity aggravated the instabilities in an already cyclical industry. As a result, many North American firms have moved to minimize their market pulp operations and concentrated instead on a strategic shift into higher value paper segments.

NEWSPRINT

This area of Canadian business strength had already attained oligopoly dimensions by the late 1920s. A tremendous investment boom followed the elimination of the US newsprint tariff in 1913. During the decade following World War I, capital investment tripled and capacity more than doubled. Over this period, the International Paper Company was the undisputed price leader in the newsprint sector east of the Rockies, and its announced prices served as the benchmark for all major contracts. Many other newsprint producers came together in joint sales companies in efforts "to control production and prevent price-cutting" (Guthrie 1941, 63). The first of these, the Canadian Newsprint Company, representing three major producers, failed within a year of its origin in 1928. Its successor, the Newsprint Institute of Canada, lasted for just under three years before collapsing markets eroded the will to act collectively. In 1933 the Newsprint Export Manufacturers Association of Canada, backed by the governments of Quebec and Ontario, hammered out a floor-price agreement together with the reallocation of a "pool" of newsprint orders from big companies to small.

In the contemporary period it has been argued that price leadership in newsprint is not so much collusive as it is barometric, a situation in which leaders "are occasionally ignored, they lose their leadership position and they actually only have the power to alter prices in recognition of real market changes" (Schwindt 1977, 97). MacMillan Bloedel used to act as a leader on the west coast, while Abitibi-Stone shares this role with several US giants in the east. Despite its strong place in the Canadian industry, newsprint has experienced major transformations since 1980. The newer TMP and CTMP processes offer growing cost advantages over the old chemical/groundwood mixtures, and twin-wire paper formers allow faster speed production. Woodbridge Reed has argued that, in the rapidly restructuring newsprint sector, there will be potential for very large mills and small specialty mills but not for the medium-sized commodity mill that was the Canadian norm (Woodbridge Reed 1988, v.3).

PRINTING AND WRITING PAPERS

By contrast to the traded sectors above, the printing and writing papers sector has continued to be protected by tariffs until very recently. It was to this printing and writing segment that American fine paper producers shifted in 1913, followed shortly thereafter by the Canadian segment. Behind the tariff wall, Canadian producers indulged freely in collusive behaviour and price-fixing, which triggered a series of anti-combines investigations and prosecutions (Skeoch 1966). In one particularly prominent case, seven fine paper manufacturers, including Howard Smith Paper Mills, the E.B. Eddy Co., Rolland Paper, and others, were convicted, along with 21 paper merchants, of unduly lessening or preventing competition during the 1933-52 period (Regina v. Howard Smith Paper Mills Ltd., *et al.* 1954, 1955). It is a telling paradox of anti-combines policy, as it has been applied to the paper sector, that activities routinely tolerated and even state-sanctioned in export businesses are subject to criminal prosecution in the domestic setting.

Only during the last decade has tariff protection begun to decline (it will be completely eliminated by 2005), presenting the Canadian printing and writing manufacturers with both challenges and opportunities. Proximity to the world's largest printing paper market, at a time when new higher value products are booming in the print advertising (quality catalogue) and in the office paper (computer and copier) fields, offers attractive possibilities. The challenge has been to identify high-growth segments on the printing product spectrum, as illustrated in Figure 2-3. Initially, Canadian interest focussed disproportionately on the foot of the value ladder, in products such as uncoated mechanical papers. These were the closest, in function, to newsprint technologies and offered the easiest path for conversion of paper lines. In time, however, the crowded nature of this newly attractive segment saw many domestic firms shifting into the higher value lines of coated papers (Woodbridge Reed 1988, v.3). The two firms analyzed below are cases in point. Domtar spun off its newsprint units in the 1990s and made a bold new acquisition in uncoated free-sheet paper for copy and office document stock as well as offset printing and book publishing. As a result, Domtar rose to number four in the world. Stora Port Hawkesbury closed its market pulp mill and reassessed the future of its newsprint mill, while diversifying into supercalendered (magazine) stock.

TISSUE

The tissue field is characterized by stable growth, fewer cycles, and higher profit margins in relation to other segments (Shahery 2002). It can be divided into a home or domestic product division—towels, tissues, and disposable sanitary products—and a commercial division—restaurants and offices. This segment tends to be the most tightly concentrated in the entire industry. In the US, Georgia-Pacific, Procter and Gamble, and Kimberly-Clark share almost 75 per cent of the tissue market. The picture in Canada is not dissimilar, with a CR_4 coefficient in the 1990s of 76 per cent. The Canadian leader is Scott Paper (33 per cent), followed by Dominion Cellulose (16 per cent), E.B. Eddy (15 per cent), and Kimberly-Clark (12 per cent). In sharp contrast to the traded character of market pulp and newsprint, tissue exports are a relatively modest component, due largely to the high transport and distribution costs associated with the product. Consequently, mills are located close to large urban markets in British Columbia, Ontario, and Quebec. Of all paper segments, tissue stands out as a manufacturer of consumer products and is highly advertising-sensitive as a result. There is also a significant potential for the use of recycled paper, which reduces the cost of fibre inputs, and product

Figure 2-3

Printing Papers—Spectrum of Products

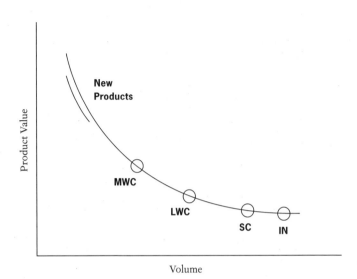

	CHARACTERISTICS	APPLICATIONS
IN: Improved Newsprint Uncoated grade Groundwood	High bulk, low weight	Limited for fine detail
SC: Supercalendered Uncoated grade	Bright, smooth, glossy	Mass circulation and full colour magazines
LWC: Light-Weight Coated Coated grade	High strength	Catalogues and magazines with advertising
MWC: Medium-Weight Coated Medium thick coated grade Groundwood	Smooth surface, high gloss printing	Specialty magazines and quality advertising

Gsm = grams per square metre

Source: Adapted from "Papermart Paper Grades" (www.papermartuk.co.uk).

differentiation is a key characteristic. While overall demand for tissue-grade products is relatively inelastic, there is potential for brand substitution, making advertising strategy an integral part of the business. The fact that tissue makers are so consumer-conscious has made brand names a key dimension of marketing—and made manufacturers extremely sensitive to public sentiment in "green" marketing areas. In a related vein, the manufacturing technology is highly proprietorial, imposing another significant barrier to new entrants.

Strategies

*Domtar Inc.: A Diversified Eastern Canadian
Pulp and Paper Producer*

In 2001, Domtar Inc. ranked second among major Canadian forest products firms. At more than C$2.5b, its sales were two-thirds the magnitude of the industry leader, Abitibi-Consolidated, but slightly ahead of third-place Cascades. Where Abitibi's strength is in the newsprint segment, Domtar sells a more diversified range of forest products, a market strategy achieved over the last half-century in response to the priorities of a series of different owners. This analysis will pay particular attention to the way Domtar's successive controlling shareholders have defined their business and political interests. Simply put, ownership matters, and each shift of control has the potential to change a firm's business trajectory. For Domtar, these blocs have included foreign investors from the United Kingdom, central Canadian syndicates, a conglomerate holding company, and a Quebec para-statal pension fund. The key points of structural transition are illustrated in Figure 2-4 below.

Surprisingly, Domtar's origin had little to do with forest products. It began as an English-owned coal tar distillery operating at Sydney, Nova Scotia. Coal tar was a preservative material used widely in construction products. By 1929 this company had expanded into a network of six tar distilleries and ten creosoting plants at locations across Canada. That year, a group of central Canadian capitalists, including Sir Herbert Holt and J.H. Gundy, took control of the firm under the new name of Dominion Tar and Chemical (Domtar). Though it was paralyzed during the depths of the Depression (1932-36), Domtar pursued an aggressive program of takeovers and expansions later in the decade, often by means of share purchases or exchanges. This included expansions into the related areas of roofing materials, fibre conduits, chemicals, and salt (Savage 1977).

In the early postwar years, the Toronto-based Argus group came on the scene. Argus Corporation was a diversified holding company that Peter Newman described as a central pillar of the Canadian business establishment (Newman 1975). Argus targeted Domtar as a prospective acquisition, and by 1950 it had achieved a controlling position. This marked a watershed for the company. Following its usual custom of assuming an active managerial role, Argus installed several key figures on the executive committee of Domtar's board of directors. Both E.P. Taylor and Eric Phillips played key roles in overseeing the company's record profits during the 1950s. Then, in

Figure 2-4

Four Production Regimes at Domtar Inc.

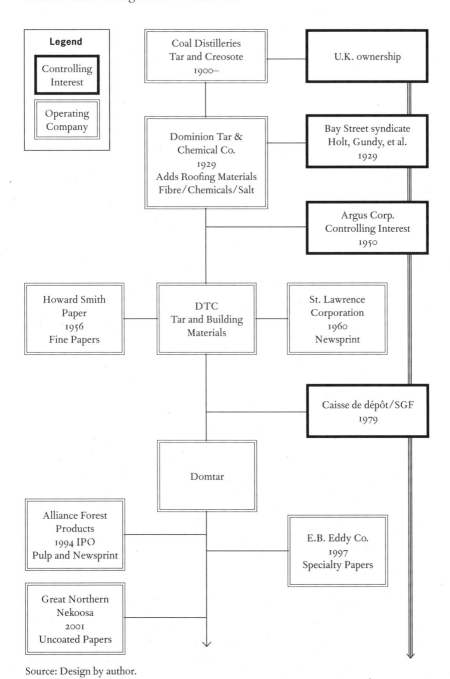

Source: Design by author.

1956, Argus launched Domtar on a dramatic and novel expansion into forest products. Typically for Argus, this was achieved by buying control of several existing enterprises. The first target was Howard Smith Paper Mills Ltd., Canada's largest manufacturer of fine and specialty papers. Argus purchased an initial 16 per cent of Smith shares on the open market, bought a second, private block to reach 33 per cent, and emerged with absolute control at 53 per cent following a Domtar-Smith share swap offer in 1957. (In a share swap, target company shareholders have the chance to exchange their shares for those in the predator firm at designated ratios.) At a stroke, pulp and paper had become Domtar's largest revenue stream.

Four years later this was repeated in the takeover of St. Lawrence Corporation Ltd., one of the country's largest newsprint producers. After assembling a 22 per cent share base by market purchases, Argus again proposed a share swap between the two operating firms. This was almost universally accepted, leaving Domtar with 93 per cent of the available shares. In 1961, St. Lawrence operated several large pulp and newsprint mills at Trois-Rivières and Dolbeau, Quebec, and Red Rock, Ontario, as well as a clutch of smaller kraft and groundwood pulp mills. Of added significance, St. Lawrence held crown timber rights to more than 9,000 square miles of crown forest in Quebec and Ontario.

Over this five-year period, Domtar's total net assets increased by 500 per cent, with pulp and paper now accounting for almost two-thirds of total revenues. Major capital improvements were made upon this base, with several new paper machines and one new kraft pulp mill installed by 1967. With Taylor's retirement from the board of directors, it fell to the new chairman, T.N. Beaupré, to consolidate and pare back the spoils of expansion. For example, Domtar closed four of its newsprint machines at Trois-Rivières in 1972, selling those assets to Kruger Inc. the following year. At the same time, the remaining two paper machines were converted to coated papermaking. This sort of conversion not only divided up a high-cost, aging facility, but also helped to balance out Domtar's product mix. However, even the Argus drive for expansion had its limits. When Quebec rival Price Brothers Ltd. became a takeover target in 1974, Domtar briefly considered extending its shareholding in Price (inherited through the St. Lawrence deal) before selling its stake to the eventual victor, Abitibi, for a net gain.

By mid-decade, Domtar's profits stood at record levels. It was solidly identified as a pulp and paper producer, holding approximately 6 per cent of national paper output. This was split between newsprint and pulp (40 per cent), which were chiefly export commodities, and other fine and coated paper grades (60 per cent), which were sold primarily within Canada. Consequently, Domtar's share of the domestic paper *market* was considerably larger than 6 per cent. As in any company, there was a strategic tension between Domtar's short-term profitability and its long-term deployment of assets (Savage 1977).

An additional intervening factor was the declining commitment of the Argus Corporation. As its principal partners aged, Argus placed more emphasis on dividend earnings than new share purchases. By 1978 its Domtar shareholding had declined to only 19 per cent as a result of share dilution through new issues. This left the senior Domtar managers exercising considerable autonomy, compared to their position in earlier decades. The internal politics of Argus Corp. took a dramatic turn in 1978, when Conrad Black acquired the controlling interest. The son of a first-generation Argus partner, Black faced mounting demands for operating revenues and chose to sell his Domtar equity to raise a stake.

This triggered a dramatic and highly public battle for control, which passed eventually to a surprising new source. When the Domtar board became aware of Black's intent to sell his share bloc to the west-coast forest giant MacMillan Bloedel in a private deal, the directors retaliated by launching a takeover bid for that company. Once it became clear, early in 1979, that the BC provincial government would not tolerate eastern control of its largest independent forest firm, both Domtar and MacMillan Bloedel withdrew their share tender offers for one another. Six months later, MacMillan Bloedel sold its initial Domtar stock, at purchase price, to the Quebec provincial pension manager, the Caisse de dépôt et placement du Québec. As part of its fund management mandate, the Caisse was empowered to acquire minority equity holdings, particularly in firms with strategic significance to Quebec. These shares were generally held passively by the Caisse, which was usually content with a seat on the board of each major portfolio company.

For Domtar, however, this marked more of a beginning than an end, as many huge forest companies were bought and sold in the ensuing years. Indeed, by the spring of 1981, the financial press was openly speculating on Domtar's status as one of the few forest giants still available for acquisition, one with an undervalued stock as well as cash on hand. With the pro-sovereignty Parti Québécois in power, the Quebec provincial government was no more willing than British Columbia to risk the loss of an "anchor" firm. In August the Caisse acted together with the Quebec state finance vehicle, the Société générale de financement (SGF), to acquire a joint 42 per cent of Domtar's equity. This signalled a more aggressive turn in Caisse strategy and was matched by a call for greater representation on the Domtar board. It also inaugurated a period when political considerations began to play a continuing role at senior levels of the firm. The Caisse/SGF move was likely triggered by Domtar's decision to shift the headquarters of its Sifto Salt company from Montreal to Toronto, together with its unwillingness to join in a new salt venture in Quebec's Magdalen Islands. Both moves were known to irritate the Quebec provincial treasurer, Jacques Parizeau, who no doubt viewed them as signals of Domtar's fading commitment to Quebec, where its forest reach was vast.

Without question, Domtar shared fully in the investment boom of the late 1980s. During the 1985-91 period, the company spent C$2.3b on expansion. Employment hit a new high of over 14,000 during this time. While Domtar's heavy debt load may have seemed manageable while the business cycle was cresting, it also held ominous implications for the next downturn.

When it came, the 1990-94 slump in the pulp and paper sector was both prolonged and severe. Domtar's experience reflected the fate of "big paper" in general, though it was qualified by certain "signature" characteristics particular to the firm. From a modest C$33m profit in 1989, the company slid into four years of consecutive loss. Even before the losses accelerated late in 1990, the board of directors had initiated an organizational shakeup. This included the termination of Chairman James Smith and several vice-presidents, with record-setting compensation of C$1.4m. The new chairman, Jean Campeau, who was recruited from the Caisse, and the new president, Pierre Desjardins, formerly of Labatt Canada, were installed in 1991. In its desperate attempts to cut operating costs, Domtar laid off more than 500 millworkers and an even larger number of salaried office staff in 1991. It closed several of its smaller, marginal mills and asked its unions for wage concessions, cutting its annual loss in half.

With no signs of recovery, a year later Domtar was in serious need of refinancing. This led to a C$600m package of new equity, notes, debt conversions, and extended lines of credit. In return for the credit, Domtar's banks imposed tight restrictions on dividend payouts, debt ceilings, capital spending, and new acquisitions. Without question, the firm was imposing system-wide disciplines throughout this period, though the effects often seemed to pale against the scale of continuing losses. Desjardins declared in December 1992 that the company was "back from the abyss," having achieved an average productivity increase of 15 per cent at the same time that the labour force was cut by 25 per cent (Melnbardis 1992). Yet, the fact remained that all four operating divisions—fine paper, construction materials, pulp and newsprint, and packaging—recorded losses on the year. Moreover, turbulence continued at senior management levels. Jean Campeau was obliged to take leave from his position as chairman in June 1992, after certain of his comments favouring Quebec independence angered major shareholders and clients.

Domtar already had in hand a consultant's report, mapping out a "spin-off and concentration" strategy for survival in the 1990s. This was aimed at bringing its debt-to-equity ratio from the prevailing 55 per cent down to a more acceptable 35 per cent. With only the fine paper plants in Cornwall, Ontario, and Windsor, Quebec, reserved as an untouchable "core," this strategy ran squarely against the diversification drive of previous decades, which had left the company with as many as ten operating divisions. One tactic was to position non-core operations for possible sale. For example, a new division combined Domtar's sawmills with its logging and forestry services. Another division, the paper distribution business known as Domtar Merchants, was merged with a larger US distribution network. In a third case, an entirely new subsidiary, known as Alliance Forest Products Inc., took control of the Dolbeare and Donnacona groundwood and newsprint mills and was financed by a new public share issue in the spring of 1994. Similar plans were laid for the Construction Materials Group and Techni-Therm, the roofing and insulation group, as well as the Quevillon and Red Rock mills. In fact, by the spring of 1994, this tactic had become commonplace in the forest sector, with many firms opting for initial public offerings to finance arm's-length subsidiaries while getting them off the parent's books.

However, the Desjardins plan was itself a victim of changing times by the fall of 1994. With the Parti Québécois back in power in Quebec City, and the Caisse/SGF holding a dominant equity bloc, Domtar's executive was under pressure. The chairman, Paul Gobeil, was a former Liberal cabinet minister, while Pierre Desjardins had organized Liberal Premier Robert Bourassa's return to politics in 1983. Both men were terminated by the board in October. However, the strategy of selling non-core assets continued. This, coupled with the pulp and paper market rebound of 1995, left Domtar with an impressive cash surplus, reduced debt, and a core of high-potential assets by the time that Raymond Royer arrived from Bombardier Inc., as president and CEO. Following the turbulent partisanship in the executive suite of the early 1990s, the situation seemed to calm during the Royer years. While the Caisse and SGF continued to vote their shares (15 per cent and 20 per cent, respectively) in selecting Domtar directors, the company's strategy took a decisive new turn in 2001.

In the meantime, Domtar concentrated on strengthening its core competencies in market pulp, paper, packaging, and wood products. A firm-wide drive for permanent cost reductions delivered more than $100m in economies, equivalent to $1 per share. Then in 1997 a joint venture with Cascades created Norampac as Canada's leading container and corrugated board firm and North America's number ten. One year later, the acquisition of fine paper producer E.B. Eddy brought an important balance to Domtar's commodity-centred paper business. However, it was the August 2001 purchase of four major US mill complexes from Georgia-Pacific Corporation that marked the emergence of the "new Domtar." First of all, it established the company as a genuinely continental producer, with 80 per cent of its sales in the US. Second, it doubled Domtar's capacity in the coated and uncoated free-sheet segments, adding particular strength in business papers (copies and documents) and printing papers (commercial, books, and magazines). Third, many of these mills were on the competitive cutting edge. The Ashdown, Arkansas, facility was described as one of the largest and lowest-cost producers on the continent. Overall, the 2001 deal catapulted Domtar to an unprecedented level, enabling it to "gain critical mass in a rapidly consolidating industry" and emerging as the world's fourth largest supplier of uncoated free-sheet (Domtar 2001). It was unwilling to be left behind as pulp and paper globalization moved to the next level.

Stora Forest Industries: A Foreign Subsidiary in a Distant Market

Historically, Canadian ownership has been relatively high in the domestic pulp and paper industry. However, US and British capital have also played a prominent role, as firms like International Paper, Kimberly-Clark, Bowater, Reed, and others have found that Canadian timberlands offer profitable platforms for international production. Many major US newspapers also took equity positions in Canadian pulp and paper plants to secure their long-term access to newsprint supplies. The first major foreign investment wave came with the resource boom of the 1920s. A second occurred in the 1960s, with European firms now involved as well. This time, projects were encouraged by provincial authorities dangling tax and grant subsidies, which have been described as tools of "forced growth"—investment at sites that would not have been chosen by market factors alone. Normally, the firms that benefited from these incentives drove hard bargains with host states (Mathias 1971).

Given this long history, the foreign-owned subsidiary firm or branch plant is an organizational form of considerable interest. Does it differ in significant ways from a domestically owned facility? Are the internal politics of the foreign firm unique? What are the policy control mechanisms within a foreign-owned subsidiary and how might these differ from a domestically owned enterprise? For almost 20 years, from 1965 to 1985, state policy for the regulation of "foreign ownership" was central to Canadian industrial policy. In the face of Canada's uniquely high level of foreign-owned manufacturing, economic nationalists raised urgent questions about the loss of business sovereignty, the inferior performance of truncated branch plants, and the dangers of de-industrialization during cyclical downturns. The government of Canada responded in several ways. Key sectors such as banking, communications, and transportation were excluded legislatively from foreign equity control; crown corporations, such as the Canada Development Corporation and Petro-Canada, were established to repatriate ownership of "strategic" firms or to fill entrepreneurial gaps; and regulatory instruments, such as the Foreign Investment Review Agency, were authorized to screen major foreign takeovers as well as major greenfield foreign investments (Canadian Forum 1971). Not all state authorities shared Ottawa's perspective, however, and most provincial governments remained staunch champions of foreign capital inflows for major project development.

The nationalist paradigm went into rapid decline in the mid-1980s, as neoliberal values infiltrated the policy establishment. The logic of continental free trade implied a comprehensive rationalization of industrial production in Canada along with an open border for capital flows. In 1985, the federal outlook shifted from foreign investment screening to foreign investment promotion. US commentator Robert Reich questioned the underlying connection between ownership and economic sovereignty by asking "Who is Us?" He queried what sort of enterprise is desired: the domestic-owned firm of possibly failing productivity or the foreign firm whose cutting-edge facilities create good jobs with long-term prospects (Reich 1990).

In many respects, the company profiled in this section mirrors both sides of the question. SFI is the Canadian subsidiary of the Swedish giant, Stora Kopparberg, now called Stora Enso. Established in eastern Nova Scotia in the early 1960s, it was part of the postwar "second wave" of Canadian pulp and paper expansion. Over the past 40 years, SFI has evolved through several organizational phases and name changes, which are summarized in Figure 2-5 below.

With integrated mill facilities and backward linkages to timber management, harvesting, and transport, SFI has been a corporate anchor in an economically depressed region. Yet at least three times in its life, the Nova Scotia company faced the prospect of closure, at moments when commodity markets, plant configurations, and multinational parent priorities seemed fatally ill-matched. The penultimate crisis in the mid-1990s was resolved by a major

Figure 2-5

Three Production Regimes at Stora Port Hawkesbury

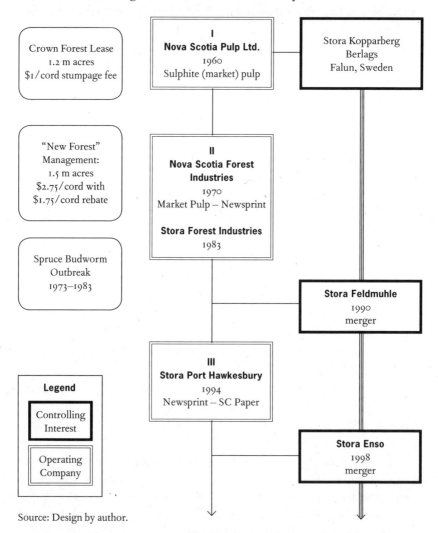

Source: Design by author.

restructuring and expansion, which transformed the mill from a structural laggard in market pulp and newsprint to an industry leader in supercalendered paper. This is the textbook transition, prescribed by strategic consultants such as Woodbridge Reed, to secure enterprise renewal and growth. However, each of these pivotal moments entailed a choice among alternative outcomes, including corporate closure and flight to offshore sites. The politics of these junctures is explored below.

EVOLUTION OF THE SFI MILL IN NOVA SCOTIA

The parent firm, Stora Kopparberg, claims to be one of the oldest corporations in the world. With business records stretching back to 1288, Stora took its name from its mining business as the "Great Copper Mountain" (Hallvarsson 1987). The company's global headquarters remain in Falun, Sweden, close by massive copper deposits. For centuries this mine generated a crucial Swedish export commodity, accounting at peak production for 60 per cent of national wealth in 1650. In the nineteenth century, Stora diversified into iron and steel and forest products, playing an integral part in Sweden's industrial revolution. Stora's forest enterprise began with sawmills in the 1870s and then diversified into pulp and newsprint in the 1890s. In 1916, control of the company passed to the Wallenberg family, which nurtured Stora as a flagship holding.

Stora's entry into North America resulted from a combination of business and political contingencies. The government of Nova Scotia was actively seeking a forest industry investor for the eastern end of the province. To this end, the deputy-minister of trade and industry combed the pulp and paper convention circuit for prospects. He encountered Stora's North American sales manager, who was based at that time in New York. This led to exploratory visits to Halifax and field surveys in the eastern counties. When headquarters proved sceptical, the sales manager took a bold step on his own initiative. He joined with some Halifax partners in 1958 to incorporate Nova Scotia Pulp Limited (NSPL) and took out an option on a massive crown timber lease to 1.2 million acres. A year later, after extended lobbying within the firm, NSPL persuaded Stora Kopparberg to assume the lease and to undertake the construction of a mill at Port Hawkesbury, whose deep-water, ice-free, seaboard location offered an additional advantage to an export producer like Stora.

As in any major greenfield investment, the initial strategic design was critical. What sort of facility would Stora choose? The company's plan called, rather conservatively, for a stand-alone sulphite pulp mill with a modest capacity of 110,000 tons per year. This choice was favoured due to Stora's expertise in sulphite processes, together with the judgement that the alternative, a kraft process mill, would require nearly twice the capacity to be commercially viable. Some measure of the province's eagerness to close the deal can be seen from the terms of NSPL's lease to the crown timber. When drafting the proposed lease, the senior government advisor recommended a stumpage (rental) fee of almost $4 per cord, the prevailing fee for new leaseholds in other expanding jurisdictions like Alberta. However, this rate was altered at the cabinet table by ministerial decision (Sandberg and Clancy 2000). As a result, the lease was granted for a 50-year term at the highly concessionary stumpage rate of $1 per cord, a rate first offered to the provincial industry in 1929. Even with these significant concessions, however, the company was not immediately profitable.

During its first decade of operation between 1962 and 1971, NSPL's sole commercial product was bleached sulphite market pulp. The business results were disappointing. Although the mill reached full capacity in 1964, pulp prices lagged below expected levels. Furthermore, while raw pulpwood remained cheap, labour costs rose by 80 per cent over this period, and the mill recorded steady losses (Soyez 1988). It appeared that Stora's initial scepticism was being confirmed. A point of decision loomed as the decade closed. Clearly, the status quo was unsustainable. The parent firm could sell, close, or expand the plant. In 1971 it chose the third option, enlarging the pulp capacity and diversifying through the addition of a 180,000-tonne-per-year newsprint machine.

This marked a major turning point for the Nova Scotia plant, now known as Nova Scotia Forest Industries, or NSFI. Its far greater wood fibre needs meant that the annual allowable cut from the crown forest lease would have to be doubled. This led the company to embrace a new, intensive approach to silviculture described as "the New Forest." It combined clear-cut logging, artificial planting, and accelerated forest rotations, along with pressure on the independent logging contractors for more mechanized operations (NSFI 1972). In addition, the crown lease was expanded by adding another 300,000 acres to the timber base. There was also a new level of government-business collaboration on forest management through a series of state-financed forest improvement agreements running from 1976 to 1994.

Despite recording its first profit in 1974, the decade remained turbulent for NSFI, both politically and economically. The company resisted the efforts of small, private woodlot owners to establish a pulpwood marketing scheme along agricultural lines (Clancy 1992). It was also drawn into sustained conflict with the government of Nova Scotia over requests, ultimately denied, to authorize aerial chemical spray to contain a major spruce budworm epidemic that began in 1974 (May 1982). Yet another political dispute arose after the first OPEC oil shock of 1973-74, when the Nova Scotia Power Corporation, a provincial state enterprise at the time, first cancelled and then renegotiated its long-term electricity supply contracts with large corporate clients such as NSFI.

By the time of the 1982 recession, Stora plant management felt increasingly beleaguered. It faced pressures from volatile international commodity cycles, local environmental activists, militant private woodlot (pulpwood) suppliers, and unreliable politicians. This was captured, both in tone and in substance, by NSFI's submission to the provincial Royal Commission on Forestry (NSFI 1983), which described a firm that had been invited to Nova Scotia only to be treated poorly by its hosts. Good faith contracts, it was said, had been either broken or unilaterally altered, while necessary management tools were constrained or denied. In such circumstances, the future of the firm could only be described as cloudy. The Stora president had actually threatened to close the mill in 1976, if chemical spray permits were not granted. Neither event happened.

There was, however, a more supportive side to this business-government relationship that SFI was less inclined to advertise. The province assisted in the salvage of budworm-damaged timber and provided significant cost-shared support for intensive reforestation under a succession of federal-provincial agreements. The provincial government also discouraged and delayed the campaign by small woodlot owners to organize commodity marketing in pulpwood, which would raise the price of raw fibre. In addition, approval was granted for herbicide spray treatments on young plantations to suppress hardwood growth on the newly reforested softwood tracts. Despite such political support, the late 1980s brought the Port Hawkesbury plant to another turning point. This time, however, the future of SFI was being determined by a far wider set of commercial and political forces, rooted not only in Sweden but also in Dusseldorf, Brussels, and Helsinki.

STORA'S EVOLVING INTERNATIONAL STRATEGY

Like any local production complex, SFI's Nova Scotia mill sits at the base of a corporate hierarchy of considerable complexity. While operational authority is normally concentrated at the single plant level, strategic planning and financial controls are far more centralized. In the 1990s, this applied not only to SFI's relations within the Stora Group, but also to Stora's relations with its parent, the Wallenberg conglomerate.

Operating through a number of non-profit holding companies, the Wallenberg family has been a dominant force in Swedish business for generations. For half a century prior to his death in 1982, Marcus Wallenberg built and directed an industrial conglomerate accounting for one-third of the total share equity in Sweden. Stora enjoyed a privileged place in this portfolio, owing in part to its claim to be the oldest company in the world and in part to its steady profitability (Clancy and Sandberg 1995).

When Peter Wallenberg succeeded his father, the continuing viability of the group was uncertain. He was forced to assume a defensive posture through a series of clashes with rival Swedish titans, including several hostile ("greenmail") challenges for control of Wallenberg assets. This necessitated huge pay-outs to buy off the shares of the corporate raiders. However, once the Stockholm stock market began to soar in 1984, much of this pressure relaxed. Indeed, Stora embarked on the greatest growth wave in its history. However, the 1990s saw renewed challenges to the Wallenberg Group. Loan losses at its flagship bank threatened to push it into bankruptcy in 1991, and huge debts resulted from the assumption of 100 per cent control of the Saab-Scania auto and aircraft consortium. There was considerable pressure to raise additional capital by disposing of assets. Although Stora was generally regarded as one of the untouchable "crown jewels," there was no question that business difficulties originating at the Wallenberg Group formed a major part of Stora's, and SFI's, operating environment during those years.

Stora became one of the world's most dynamic forest companies after 1984 through a series of carefully calibrated mergers and takeovers. Newly appointed President Bo Bergren was the driving force in this strategy of forward integration along the paper products chain. In 1986, Stora took control of Swedish packaging and board manufacturer Billerud. The following year, the target was fine paper producer Papyrus. In 1998, Stora spent just under $1b for full control of Swedish Match. This resulted in Stora becoming Sweden's largest, and Europe's second largest, forest firm. From a total workforce of 9,000 in 1984, the company had grown to 55,000 by 1989.

Late in 1988, Bo Bergren underscored his company's new orientation with the comment that "our future lies in European integration" (Greenspon 1988). This was confirmed in the spring of 1990 with Stora's spectacular takeover of German paper giant Feldmuhle for US$2.5b. Whereas the earlier deals were designed to extend Stora's product range, this one aimed to integrate Stora's core strengths in market pulp and newsprint with Feldmuhle's papermaking capacity inside the European Union. (Sweden did not join the European Union until 1995.) As the largest producer in Europe, Stora now held strong positions in each major product area: market pulp, newsprint, light-weight and coated papers, paperboard and box, and specialty papers (Price 1991).

One of Bergren's famous aphorisms was that "contrary to what they write in the personal advice columns, size is enormously important" (Hallvarssen 1987). In 1991, after five years of strenuous expansion, Stora ranked fourth among global forest corporations. In the process, however, it had taken on a considerably enlarged debt load, just in time to witness the decline of key commodity markets in the face of recession. In this context of tight markets, system-wide rationalization, and a new European focus, the future of the relatively small "orphan" plant in Nova Scotia was open to question. Prior to the expansion, SFI figured as one of Stora's few non-Swedish ventures, and the Nova Scotia plant operated as one of six divisions of the parent company. This arrangement afforded Port Hawkesbury managers direct access to senior corporate officials, even though Stora's Swedish mills dwarfed SFI in both size and strategic importance. By 1992 the situation had changed. While there was no reason that a deeper commitment to Europe ruled out modernization in North America, SFI's continuing fit was far from obvious. In this period of retrenchment, the scenarios ranged once again from closure to sale to expansion. This was exacerbated by the fact that all Stora plants were being compared against increasingly rigorous standards of "big plant" efficiency. Equally, in competing for major capital inputs, SFI now faced a far wider set of rival sites.

Over its 40 years in Nova Scotia, Stora has faced three points of crisis. The first occurred in the late 1960s, when insufficient capacity and weak markets, together with higher than expected operating costs, raised the spectre of withdrawal. The second occurred in the late 1970s, when sudden timber losses and spiking energy costs called the plant into question once again. The third emerged in 1993, when the combined pressures of slumping markets and company-wide rationalization were brought to a head by impending federal mill-effluent regulations. Note that, in each case, the crisis emanated from imperatives within the Stora/Wallenberg circle. It was never a question of whether pulp and papermaking was viable in eastern Nova Scotia, but whether the prevailing configuration was optimal within the corporate group. Clearly the business economics of large investments are defined within the context of firms, not just of markets.

THE CLOSURE/RENEWAL CONTROVERSY OF 1993-1995

In August 1993, a senior Stora officer travelled to Halifax to inform Nova Scotia Premier John Savage that the continued operation of the Port Hawkesbury mill would be reviewed at an upcoming global board meeting in Sweden (Scott 1993). The premier was told that the SFI plant had been losing money since the onset of the 1990 recession and that no recovery was expected for at least a year and a half. Moreover, SFI faced an imminent decision on a major, $65m investment for new effluent treatment facilities, which were required to meet new federal emission standards coming into effect at the end of 1995. These combined pressures forced Stora to make a strategic choice. Since the mill complex was judged not viable in its present configuration, it could either be closed, thereby saving the $65m along with ongoing operating losses, or reconfigured and expanded to become profitable again. The latter option was deemed to require new newsprint machines, as well as a de-inking and recycling mill to meet the growing US state regulations on recycled fibre. In effect, the board would be deciding whether a capital investment of close to $700m should be committed to Nova Scotia or to alternative locations within Stora's corporate empire.

News of the premier's meeting soon became public, triggering a wave of panic throughout eastern Nova Scotia, where SFI was the largest private-sector employer. The company claimed to provide 1,200 direct jobs, together with a multiplier effect generating two to three times that number indirectly. Regardless of exact numbers, one local resident summed up the stakes, noting that there is scarcely a home of more than 1,200 square feet that is not connected in some way to the mill. The Port Hawkesbury mayor spelled out the political corollary by declaring that "there's God and there's Stora Forest Industries" (CBC 1993).

For the provincial government, this news could not have come at a worse time. Only three months in power, the Liberal administration was grappling with a severe fiscal crisis, highlighted by an inherited deficit of close to $600m (Clancy, *et al.* 2000). In an effort to reassure provincial creditors and bond-rating agencies, the finance department had been struggling to prune this figure by at least $200m. Put simply, the provincial government lacked the fiscal capacity to tip the scales with subventions toward an SFI modernization. At the same time, the government had made rural economic development one of its central political themes, and SFI's demise would be a massive setback.

A "grassroots" rescue campaign was orchestrated by elements of the business network most closely connected to the pulp and paper mill. To use Baron's (1999) terminology, a political coalition was constructed along the Stora rent chain. Virtually from the moment of the August 1993 announcement, SFI's Nova Scotia president and general manager had argued that the plant's fate hinged on a revised business strategy that could be put to the Stora board in Sweden during its review. To this end, plant management opened negotiations with various business interests within its orbit, seeking to reduce the mill's operating costs. Initially, a proposed 5 per cent wage cut and two-year wage freeze were rejected by Local 972 of the Communication, Energy and Paperworkers Union (CEP). A second, lesser wage freeze was approved four months later. On the other hand, talks with the major private wood suppliers association and the pulpwood truckers, both made up of small independent business interests woven tightly into SFI's product chain, were more successful. Both pulpwood prices and truck hauling fees were reduced on a multi-year basis. The local municipality offered property tax reductions, and the mill management held discussions with rail and electric utilities, though no firm agreements were publicly announced.

A number of local support committees were also struck during the fall of 1993. The Port Hawkesbury Chamber of Commerce launched a lobby campaign with provincial and national politicians, underlining the crucial economic importance of the mill and challenging the need for more stringent effluent standards. A second group, formed by regional politicians from the seven eastern Nova Scotia counties, sought an extension of the 1995 federal regulatory deadline. For its part, the provincial government offered moral support but refused to endorse a regulatory exemption, appreciating no doubt that any concession to SFI would be difficult to deny to the other four major forest companies. However, as the year drew to a close, the government announced a $15.4m forgivable loan to SFI, ostensibly to support woodlands improvement. Neither the basis of this loan nor its significance to SFI was ever made clear.

By the end of 1993, the Stora board had postponed its critical meeting in order to take account of changing circumstances. No doubt the incremental concessions at SFI had been noted in Sweden, though these could not address the structural weaknesses already noted. More significant surely were signs (illustrated in Figure 2-1) that market pulp and paper prices had begun to recover from their 1992 trough. In Nova Scotia, the SFI manager suggested that the newsprint expansion with recycling facilities could stand alone, while the market pulp line could be closed. The first sign of new possibilities came in February 1994, when the Stora board authorized $36.5m for a secondary treatment facility for the newsprint mill alone. The market pulp operation was guaranteed only until the end of 1995, something which the Nova Scotia government now claimed had been a precondition for its "loan." At the same time, SFI reiterated its call for an extension of the 1995 effluent compliance deadline. As 1994 unfolded, the market pulp sector staged a strong recovery, with prices rising from the $500 per tonne range to over $800 per tonne. In August, Stora authorized a larger scale treatment facility (now $48m) to cover pulp mill effluents as well. Thus, the short-term prospects for the SFI mill had unquestionably improved, though the long-term picture remained in flux. Indeed, a Swedish trade union leader who sat on the Stora board cautioned that the SFI newsprint mill was secure for the duration of the current business cycle at best. Despite the events described above, there remained doubt as to whether any set of policy concessions could guarantee the SFI mill for the long run.

The SFI review was completed in 1995. To the vast relief of eastern Nova Scotia, Stora announced plans for enterprise renewal by building a new coated paper mill at its Port Hawkesbury complex. Opened in 1998, this new super-calendered mill is one of the most advanced of its type in Canada. In strategic terms, Stora Port Hawkesbury shifted several steps up the value-added ladder outlined in Figure 2-3, from a market pulp/newsprint configuration to newsprint/coated paper. The chemical pulp mill was phased out, and the plant's entire groundwood pulp output was captured in forward manufacturing. It appeared that the SFI complex had shifted from being one of the most marginal and vulnerable to one of the most competitive and secure operations in eastern Canada. Just months before the opening of the new paper plant, the mega-merger was announced between Stora and Enzo-Guzeit (referred to hereafter as Stora Enso).

A CHOICE OF FUTURES, 2002

In the summer of 2000, the situation took another turn when Stora Enso acquired the US-based Consolidated Papers Inc. This consisted of a package of seven mills, located in Wisconsin and Minnesota, which manufactured coated and calendered papers of varying weights along with packaging and labelling products. At a stroke, Stora Enso became one of the leading North American producers of these commodities, with facilities close to the major midwest urban markets. The structure of this division, along with its place in the wider corporation, is evident in Figure 2-6 below.

It was to this network that the Port Hawkesbury mill was joined by the establishment of a new North American division headquartered in Wisconsin Rapids. The prospective "fit" between the Nova Scotia and Wisconsin facilities was mixed. On one hand, Port Hawkesbury's new paper machine, PM-2, was the newest and largest producer of supercalendered paper on the continent, having set repeated world speed records in its early years of operation. Not only did it mesh closely with the printing, magazine, and specialty stock from the former Consolidated mills, but it also enjoyed ready access to Atlantic seaboard markets that would complement rather than duplicate Wisconsin sales. On the other hand, Port Hawkesbury's aging newsprint line had no counterpart in the US operations. For this reason, its future was far less assured.

The question became public in March 2002 when Port Hawkesbury announced that it would oppose regulatory applications by the provincial energy utility for electricity rate increases. In fact, mill officials called for the reduction of power rates, said to account for 20 per cent of production costs, to secure the future prospects of the newsprint mill (Myrden 2002). This proved to the opening shot in a wider campaign to reduce wood fibre and labour costs

Figure 2-6

Stora Enso Corporate Organization and Port Hawkesbury Mill

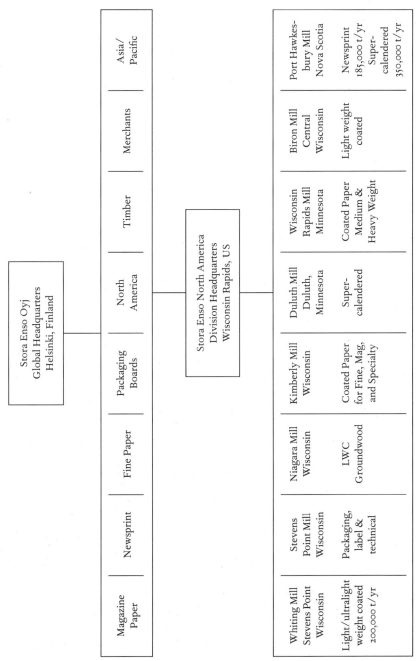

Source: Design by author.

as well. By May, mill management had requested CEP Local 972 to reopen its contract in order to renegotiate flexible work rules and asked the private wood suppliers association to cut the cost of delivered pulpwood. American officials described the mill's status as "critical," and the need for it to meet global standards was stressed (*Chronicle-Herald*, 2002). In June, with labour talks stalled, a formal date of closure of the newsprint mill was set for 31 July. Only in the dying days of July, with the preparations for closure underway, did the paperworkers agree to new work practices. In the meantime, the fibre suppliers agreed to reopen their contract and cut wood prices.

Although these concessions had been advanced by Stora Enso manage-ment as a "survival plan" for the newsprint mill, the exact level of security remained unclear. For one thing, the question of electricity costs remained to be resolved. Though Nova Scotia Power's 2002 application for a rate increase had been denied, Stora Enso continued to press the power corporation for a special high-user rate. A bare six months later, Port Hawkesbury stated once again that, without an acceptable power deal, the newsprint mill and a possible $100m thermo-mechanical pulp plant expansion would be lost. Entering 2003, the future configuration of the mill was as uncertain as it had been 12 months earlier. Only after the provincial regulator had approved the lower electricity rate did the Stora Enso board consider, and eventually approve, the TMP/newsprint renewal (MacDonald 2003).

This points to the dilemmas faced internally, in the local plant, and ex-ternally, in the host society, by a transnational corporation bent on imposing commercial discipline based on the lowest commercial denominator. The logic of transnational management, as applied within Stora Enso's North American division, meant that Port Hawkesbury was forced to compete for investment capital and product mandate with the network of former Consolidated paper mills centred in Wisconsin. Some of these mills, such as Duluth, were non-union operations with flexible work rules that even extended between plants. Others were industry leaders in their product segments and therefore repre-sented the most rigorous of competitive benchmarks. This time the complete closure of the Port Hawkesbury facility was not an issue, as it had been in 1993 (the quality of PM2 operations ensured this). However, the corporate managers in Wisconsin Rapids were well aware that a reconfigured pulp and supercalendered paper facility could flourish without the 30-year-old news-print line and with less than half the labour force.

While public policy issues served as triggering events, the business "com-pact" that is re-opened in such cases involves far more than state policy con-cessions. In 1993, the looming deadline for pollution abatement compliance provided the hook on which the mill "crisis" was hung, while in 2002 it was the regulatory decision on electrical power rates. In each case, however, these issues were propelled by cyclical pulp and paper downturns, falling prices, and business losses. This is when multi-plant firms opt for company-wide reviews. The Stora and Stora Enso style of government relations is less unilateral than some of its rivals, who simply announce such decisions as *fait accompli*, but no less blunt in its purpose or sweeping in its potential effects.

For the network of suppliers, contractors, workers, government authorities, and social stakeholders within the Port Hawkesbury orbit, the danger is that each cyclical slump will bring a new version of the problems detailed above. This is, however, inherent in contemporary transnational operations. Each plant or establishment must simultaneously manage two political configurations: one within the global corporation and the other within the host jurisdiction. As our examples attest, this takes place in increasingly compressed time frames and within ever-shorter cycles.

Policy Issues

The Regulation of Pulp Mill Effluent

The regulation of pulp mill effluent represents the most significant issue in modern state policy for pulp and paper. Over the past 30 years, measures covering effluent regulation have impinged on the Canadian industry more than any other orders or laws. As such, they offer a number of insights into the mechanisms and consequences of business power. It is especially interesting to consider the role of uniform or horizontal regulatory policies on such a diverse and variable industry. A further complication arises from the shared environmental jurisdiction of central and provincial states, since the two levels can display markedly different policy inclinations in this field. Also, perhaps inevitably, the federal regulatory regime has been developed in close consultation with the industry interests that stand to be most closely affected. In addition, the fact that pulp mill pollution affects an interested public extending well beyond the producers themselves means that organized business must grapple with some disciplined and dedicated non-business adversaries. This was increasingly the case as first-generation effluent policies gave way to the second generation. Finally, this case illustrates the differences between policy-making and policy implementation as separate stages of the political process.

The pulp and paper industry found itself heavily exposed when awareness of industrial pollution first shot up the policy agenda in the 1960s. A watershed event was the 1962 publication of Rachel Carson's *Silent Spring*, a powerful critique of the damage inflicted by industrial chemical compounds on bird, animal, and marine food chains. Previously viewed as miracle treatments, common but powerful insecticides such as DDT were described by Carson as untested hazards to the biological web of life. Government authorities were also criticized for neglecting to defend the public interest and the natural world against the despoliation brought by narrow, profit-seeking business agendas. While the US chemical industry mounted a full-scale attack on Carson's credibility, public and media concern kept the issues alive. In Canada, the devastating impacts of mercury effluent from the Reed Paper complex at Dryden, Ontario, which came to light in the early 1970s, helped crystallize public concerns about uncontrolled toxic by-products in pulp and paper (Troyer 1977). Pollution awareness grew with each new industrial debacle, from oil tanker groundings to factory chemical dumps. By 1970, 1 May was being celebrated by citizen environmentalists as "Earth Day." Richard Nixon established the US Environmental Protection Agency, and Pierre Trudeau created Canada's first Department of Environment (DOE), with the Environmental Protection Service as its regulatory arm.

Industrial pollution control gained public prominence as part of the new style of "social" regulation that became a hallmark of the 1970s (Vogel 1986). Similar to consumer protection and workplace safety, its targets are what economists call the "externalities" of commercial production: the consequences that spill over the borders of enterprise to significantly impinge on wider public arenas. In response to serious encroachment, the state intervenes to monitor and limit externality effects by regulating select dimensions of business behaviour. Beginning in the 1970s, social regulation became a political battleground in all advanced capitalist states, as business enterprise challenged the principles, techniques, and costs of this new style of rule-making and compliance.

As the largest source of water-borne industrial pollution in Canada, the pulp and paper industry was an understandable early target. It was estimated that one-half of all industrial wastes released into water came from this sector. Inevitably, new regulations posed a major challenge to the approximately 120 mills then operating in nine provinces. Yet not all firms and plants stood to be affected similarly or equally, given the considerable variations in technical processes and plant age. Geographic location was another important consideration, particularly for water-borne effluent, given variations in the patterns of physical diffusion and social exposure. Should plants operating at tidewater be treated the same as those on inland freshwater river systems? These factors also complicated the role of trade associations in mobilizing a collective industry voice. Mills had differing priorities according to age, level of sunk capital, profitability, ownership, and production processes. The same factors confronted state authorities as they searched for regulatory responses in this highly dynamic field. Did pulp and paper, as a leading industrial export sector, merit special consideration? Should the economic and social interests of single-industry mill towns be weighed against those of wider environmental integrity and health? Which level of state held jurisdiction over industrial pollution and how should this authority be exercised? There were no obvious answers to these questions in the 1960s, when environmental control began to emerge as a crucial field of social regulation. This policy case highlights the impact of business lobbying, bureaucratic politics, federal-provincial tensions, public interest advocacy, and scientific research in the ongoing process of state regulation. To capture the shifting business and political dynamics, three phases or "generations" of policy are discussed.

THE FIRST GENERATION OF EFFLUENT REGULATION UNDER THE FISHERIES ACT

The first Canadian regulations for pulp and paper effluent appeared in 1971. Before dealing with their particular rules and their impact, it is important to appreciate the political institutional context and the policy frame that guided the debate and design on effluent control. Environmental protection is a vast field, encompassing airborne and water-borne pollution, solid waste disposal, and landscape and ecological integrity, among other matters. Within the federal system, environmental policy has evolved as a shared jurisdiction between Ottawa and the provinces, a fact that has complicated both program design and delivery. Within any single state, further complications arise from the fact that many departments and agencies tend to have responsibility for laws and regulations of an environmental sort. Since they bring distinct mandates and philosophies to bear, the challenges of achieving policy coordination and coherence are compelling.

Even though paper manufacturing entails significant airborne emissions, the manufacturing requirements of heavy water intake and output suggested an initial aquatic policy focus. From the start, the government of Canada held one powerful statutory instrument in the form of the federal Fisheries Act. More than a century old, it had long been used for point-by-point protection of key fish stocks, but it was never designed as a comprehensive instrument for the control of industrial pollution. Nevertheless, from about 1966, officials in the Department of Fisheries and Forestry (DFF) began to recognize the systemic nature of the pollution problem and the fact that, by exercising its mandate for fish and fish habitat protection, Ottawa could regulate water quality and thereby industrial effluent discharge. Over the next five years, as it haltingly explored these new directions, the DFF emerged as the *de facto* federal regulator of the paper industry. Toward the end of this period, in 1971, the federal fisheries bureaucracy was absorbed into Canada's new DOE. By then, a "memorandum of understanding" between the senior officials in Fisheries and Environment authorized the newly created Environmental Protection Service to assume regulatory responsibility for pulp and paper pollution under the Fisheries Act.

Significantly, a second approach to pollution control was also available through the recently approved Canada Waters Act of 1970. Its administrative responsibility lay with the federal Department of Energy, Mines and Resources (EMR), and the policy paradigm was based on the comprehensive physical and social analysis of entire river basins as regulatory units. Significantly, its fulcrum of regulatory control was the "assimilative capacity" of the receiving waters to absorb industrial effluent without significant biophysical damage. This implied that permissible effluent levels could vary according to site conditions, being different in vigorous tidal waters than in small freshwater lakes or slow-moving river systems. As well, a far greater degree of provincial collaboration was also anticipated in the field of water planning, with joint federal-provincial management boards being proposed as regulatory bodies. In Ottawa, an extended bureaucratic battle unfolded as the two agencies, DFF and EMR, advanced their rival strategies (Parlour 1981).

Politically, it was DFF that offered a faster and more direct route to the control of the pollution "hot spots" that were increasingly making the news. Consequently, in 1969 the cabinet approved the drafting of the necessary amendments to the Fisheries Act. With their enactment the following year, DFF turned its attention to developing the all-important pulp and paper effluent regulations, which would spell out detailed standards and procedures. A DFF task force fashioned an approach based on the specification of treatment processes according to the principle of "best practicable technology." This borrowed extensively from a US study that distinguished average mill effluent by three classes of facility: older, current, and new/next generation. When both industry and provincial government interests proved highly critical of the DFF draft proposal, a tripartite (federal-provincial-industry) task force was struck to make a new start in 1971. By stipulating certain non-negotiable principles, the federal minister of fisheries was able to lock in some basic features from the earlier stages, such as best practicable technology and uniform national standards. For its part, the industry, represented by the CPPA, secured separate treatment for older mills, at effluent levels lower than those stipulated in earlier drafts. Provincial governments also asserted their own pollution control jurisdictions.

While industrial effluent was a priority for the new DOE, which had absorbed the DFF in 1971, it also faced a major challenge in establishing an effective policy approach to pollution control. On one hand, it encountered bureaucratic rivalries with business-oriented departments such as the Department of Industry, Trade and Commerce, which had long championed competitive export industries such as pulp and paper, while resisting social regulations that significantly burdened plant cost structures (Macdonald 1991). On the other hand, DOE depended heavily on an act "belonging" traditionally to the DFF. Internally, DOE faced a major challenge in recruiting the scientific and engineering expertise required to analyze this new generation of policy problems. Well aware that its opening initiatives would define the foundations for pollution control, DOE identified four sectors for early regulation. While the list included iron and steel, petroleum, and mining, it was pulp and paper that figured as the prototype. Significantly, the chemical industry was judged too complex to tackle in the first wave and was set aside until new legislation could be devised (Doern and Conway 1994).

Selection of pulp and paper was easily justified, as it accounted for more than half of all industrial effluent discharged into Canadian waters (Webb 1988). But fashioning a pollution control regime was a different matter, considering the diversity of ages, technologies, and receiving water systems among the mills spread across the country. The Pulp and Paper Effluent Regulations (PPERs) steered a course midway between heavy coercion and voluntary compliance. Their basic principles were: 1) technology-based treatments; 2) end-of-pipe discharge criteria; 3) national minimum standards applied by mill class categories; and 4) negotiated compliance agreements for exempted facilities that failed to meet the "guidelines." While the shift from laissez-faire to regulated emissions implied major capital outlays for producing firms, the PPERs sought compromise rules, acceptable to both industry and the provinces. For example, regulation based on the best practicable technology principle was judged to be commercially feasible, whereas regulation based upon receiving water quality was deemed unattainable for the time. By focussing upon limiting external discharges, the regulations allowed plants a choice of internal industrial process controls, external effluent treatment controls, or both. By requiring full standard compliance in plants constructed after 1971, lesser standard compliance for plants that either converted to a new manufacturing process or expanded by more than 10 per cent of output, as well as voluntary guideline compliance for plants built before 1971, the regulations imposed a tiered approach that segmented the industry while settling for gradual improvements over time. Through it all, the result of the DOE plan was to install minimum national standards to preclude pollution havens, while leaving individual provinces the option of raising the regulatory bar by specifying additional criteria or levels in particular cases (Sinclair 1990).

In setting the detailed standards, the tripartite task force of technical officials from industry and government recommended numerical values for allowable discharges along three dimensions. Two core standards of water quality were designated. One pertains to total suspended solids (TSS), the particles of matter borne by liquid effluent, which settle eventually into marine environments, altering waterbed conditions and smothering habitats. Since pulp and paper solids involve cellulose fibre, wood particles, lime mud, and other residues, the treatments of choice include screens, filters, and clarifiers, which can reduce TSS levels in most mills by 80 to 90 per cent. A second water quality standard involved the biochemical oxygen demand (BOD) of effluent matter, which depletes the dissolved oxygen level in water and kills aquatic organisms. Biological treatment can take the form of oxidation processes that can reduce BOD levels by 70 to 95 per cent. Finally, the regulations addressed acutely toxic elements identified in effluent, such as chlorofluorocarbons and, more recently, dioxins and furans. Schedules attached to the November 1971 regulations set out the detailed numerical settings for all three dimensions.

The influence of the paper industry can be seen at various stages of both the regulation-making and the enforcement processes. As the collective business voice, the CPPA was active from the beginning. Part of its early strategy was to resist federal encroachment by presenting industrial pollution as a provincial responsibility (Harrison 1996a, 64). Later, when it became obvious that Ottawa was destined to play a role, the CPPA sought to soften the regulatory approach, calling for flexible standards, consultative rule-making, and cost-shared investments in pollution control (Parlour 1981). The CPPA benefited when the DFF opted for the more tangible focus on discharge standards as the locus of regulation, as opposed to the assimilative capacity approach being advanced by water managers in EMR. However, it is also likely that the indirect, bureaucratic representation of sector interests through the federal Department of Industry, Trade and Commerce, which warned of a massive threat to as many as 40 "marginal" eastern mills if the rules were too stringent, played an important role in setting the political context for drafting and approving the PPERs.

Within the DOE, it was the Environmental Protection Service, consisting of directorates for both water and air pollution control, that assumed the policy lead. Doern (1995) has noted here that the "regulatory" culture of the Environmental Protection Service ran against the "science and service" tradition of the rest of the department. During the first generation of rule-making, it relied on a task force approach in which technical groups drawn from industry and government worked toward consensual standards. The absence of vigorous environmental pressure groups facilitated this process in the early 1970s. In the actual deliberations, the government agencies found themselves alarmingly short of technical and commercial expertise. This allowed the CPPA to inject specialized data on a variety of points that went virtually unchallenged. Industry argued effectively for the acceptance of the tiered approach, with the crucial "grandfathering" provisions for pre-1971 facilities. It also supported the provinces in their campaigns for parallel regulations to be reconciled with the PPERs by federal-provincial cooperation. While provinces like Ontario set "local" standards at a level higher than Ottawa's, other provinces, particularly Quebec and British Columbia, set the bar lower (Harrison 1996b). Further yet, the CPPA argued successfully for federal TSS and BOD levels below those being applied in the US. The result was a regulatory bargain between the Canadian industry and governments, based on the "lowest common denominator" principle (Doern 1995).

Could the industry bloc have achieved a more favourable outcome? This seems unlikely, given that pollution regulation was coming quickly and that Ottawa was destined to play a leading role. Of the available paradigms, the technology-based, tiered-standard, cooperative version was highly compatible with paper industry priorities and secured business interests a more favourable position as a policy clientele. Furthermore, looking beyond the regulatory model, it is evident that the industry gained a further set of advantages at the implementation stage.

The DOE succeeded in winning political support for federal fiscal incentives toward new capital investment in pollution control assets. Since 1965, federal sales tax rebates and two-year depreciation for tax purposes had applied. Beginning in 1971, direct federal grants for innovative effluent system research followed. This marked a significant departure from the principle of "polluter pays," enunciated by DFF minister Jack Davis at the outset of the process. Second, the binding coverage of the PPERs was extended only very slowly. As late as 1988, only 40 per cent of all Canadian mills—11 new facilities and 40 altered ones—were compulsorily covered. The greatest proportion of paper mills still fell into the "older" category, where the 1971 regulations served as guidelines rather than binding rules. In such cases, state authorities could only approach those companies to negotiate voluntary strategies for the eventual attainment of the federal standards. The policy instrument to formalize such undertakings was the "compliance agreement," a sublegal expression of intent which could be monitored and reviewed over time. The difficulty stemmed from the non-enforceable nature of such deals. Operating in a legal grey zone, Ottawa had little choice in situations of non-compliance but to renegotiate the terms with the polluter and try again. There was no serious sanction against failure to measure up.

By this lenient approach Ottawa guarded against the danger of pressuring already marginal plants into closure. However, it also created *de facto* pollution havens, something which the policy aimed expressly to avoid. In 1985, a mere 18 mills, or 15 per cent, of the 122 then in operation accounted for 42 per cent of total industry TSS loadings and 50 per cent of BOD discharges, as well as 57 per cent of acute toxic effluent (Sinclair 1991). A decade earlier, in an effort to exert moral suasion for voluntary action, Minister Davis had threatened to "name and shame" the extreme polluters. However, by dramatizing the ongoing weakness of Ottawa's PPER regime, the potential political embarrassment associated with such manoeuvres, could only intensify as time passed, and the tactic was never employed.

Finally, pulp and paper producers benefited from Ottawa's loss of regulatory energy in the late 1970s. There were several reasons for this. With the provinces catching up with their own regulatory standards, the vacuum was being filled. Moreover, the federal government was pursuing a more cooperative relationship with the provinces, which it sought to demonstrate in the environmental field. Furthermore, within the DOE, now renamed Environment Canada, a new set of pollution priorities demanded attention, including acid rain and recently discovered toxic pollutants. Finally, Ottawa switched its policy style from regulatory oversight to financial incentives for enhanced pollution control. (This is discussed in the second policy case to follow.) The result was a federal regulatory hiatus of close to a decade when it came to pulp and paper pollution.

From the outset, Ottawa recognized that the 1971 standards and levels were only a starting point. Gearing the system to best practicable technologies carried the inevitable corollary that, as commercially adaptable systems improved, the policy expectations, goals, and criteria would need to be reviewed and revised. However, this crucial step proved politically impossible, and as a result the 1971 system remained frozen in place for more than 20 years. On the several occasions when Ottawa initiated tripartite reviews, consensus proved impossible as one or more parties exercised vetoes in favour of the status quo. In their defense of the lowest regulatory denominator, industry voices like the CPPA saw much to gain by maintaining the 1971 system. It became, in effect, the industry's consensus position, although it was not equally advantageous to all operators. Given that most of the older mills were located in Ontario and Quebec (18 of Ontario's 28 mills were opened before 1921, as were 26 of Quebec's 47), there was a pronounced regional skew to the benefits of exemption. By 1988, only 40 per cent of all Canadian pulp and paper plants were covered by the compulsory PPER standards as new, altered, or expanded facilities. By the 1980s, many commentators agree, Ottawa had lost both the political and administrative impetus for regulatory review (Sinclair 1990).

In effect, the 1971 system provided a flexible field. The industry lobby exerted relentless pressure for moderate to minimal standards. Meanwhile, each company or plant could pursue a commercially maximizing strategy of its own, centred on the art of managing its compliance agreements. Marginal mills prized their grandfathered status and invoked a variety of factors as a generation of agreements came and went (Sinclair 1990). The postponement of regulatory investments for a few more years represented both a commercial and a political victory. On the other hand, greenfield mills opted for state-of-the-art technology as part of their initial installations, aiming to curb most of their prospective effluent problems at the production stage. In effect, each corporate type pursued a strategy that anticipated the eventual tightening of PPER standards, though the older mills saw it coming later and the newer mills sooner. Questions can also be raised about the impact of this relaxed and increasingly outmoded effluent regime on the pace of pulp and paper restructuring in Canada (Sinclair 1991). At least two aspects are worth considering. First, did the extended grandfather provisions postpone or eliminate the pressure for firms to upgrade their capital stock in marginal plants and thereby contribute to long-run productivity losses within the industry? Second, did the free pass on pollution serve to postpone decisions to rationalize, both within multi-plant firms and between firms, at a time when the trends to greater scale and advanced process technologies were amply evident in rival pulp and paper sectors of the US and Scandinavia? In other words, did it slow the consolidation of global leaders in the Canadian industry?

For over 20 years, the control regime for pulp and paper effluent operated as a classic case of a clientelistic regulatory network. While this structure was born at a time of growing politicization, the decisive choices of policy design both in law and administration were shaped closely by industry and provincial

government interests. The lopsided information flows, the complicated tripartite consultations, and even the internal bureaucratic politics at the federal
level, served to determine this result. Once enacted, the flaws in legal standing
and the technical and economic limitations at the state level combined with an
increasingly ritualized form of policy bargaining at the firm and plant level.
In sum, the PPER regime had become a classic instance of accommodative
politics within a permissive, or weak, state regulatory structure that proved
highly resistant to change.

TOWARD SECOND-GENERATION REGULATION: CEPA LEADS THE WAY

In the mid-1980s, Ottawa showed signs of developing a new environmental
sensitivity. Several factors help explain this shift: the determination of the
Trudeau Liberals in their final term to reassert a national government presence; the vigour of a new generation of green lobby groups, particularly, in
the pulp field, Greenpeace; the rising profile of pollution hazards in public
opinion; and the accumulation of scientific evidence on new toxic substances.
These vectors culminated in the Mulroney Conservative years of government.
They are manifest in two key policy initiatives—the regulation of new toxic
substances under the new Canadian Environmental Protection Act (CEPA)
and the second generation of PPERs that were issued in 1992.

For many years, the weaknesses of the federal Environmental Contaminants
Act (1975) were widely acknowledged. Then, with the 1985 discovery of the
"toxic blob" in the St. Clair River downstream of Sarnia's petrochemical valley, it became clear that Ottawa had no effective tools for dealing with such
extreme hazards. Consequently, the CEPA was designed not only to plug these
gaps but to establish a new framework for toxic substance regulation (Doern
and Conway 1994).

Along with the chemical, petroleum, mining, and metals industries,
the pulp and paper sector had a strong stake in the CEPA deliberations. In
this, however, it was only one business interest among many. But developments during 1987 brought pulp and paper dramatically to the forefront,
reinvigorating state regulators in a context of unprecedented public concern.
This began with the discovery in pulp mill discharges of highly toxic organochlorines such as dioxins and furans. It was the 1987 announcement by the US
Environmental Protection Agency that dioxins had been detected both in fish
downstream of pulp mills and in paper products that galvanized regulatory
action. Greenpeace added urgency to the Canadian scene when it released test

results finding dioxins in sediments near the Harmac pulp mill on Vancouver Island (Macdonald 1991, 223). A series of fisheries closures followed, together with an accelerated federal government program of receiving water sampling. In turn, this triggered an industry-government task force in 1988.

At issue were the pollutant compounds created by the use of chlorine as a pulp and paper bleaching agent. As mentioned earlier in the chapter, kraft or sulphate mill methods yield a coloured pulp that must be whitened by bleaching. By the 1960s, kraft had overtaken sulphite technologies in chemical pulping, and chlor-alkali plants, which transformed salt into caustic soda and chlorine gas, supplied the key ingredients. Elemental chlorine gas (Cl_2) served as the standard industrial bleaching agent. When chlorine gas contacts organic (carbon-containing) molecules, as it does in the waste water of kraft pulp mills, they bond into organochlorines. Highly persistent, bio-accumulative, and in many cases toxic, over 300 organochlorines have been identified. While dioxins and furans were the first to be identified and regulated, they are only the tip of the iceberg.

The carcinogenic potential of dioxins and furans constituted a hazard of an entirely new order, and a vortex of political forces swirled around this question. Governments felt intense pressure to act, and, in Canada, the provinces were first out of the gate. The working standard, pioneered some years earlier in Sweden, was designated as absorbable organic halogens, or AOX, expressed as kilograms of AOX per air dried metric tonne of pulp (kg/ADt). It should be noted that the AOX standard went beyond dioxins and chlorine to measure all halogen compounds that might be bonded or absorbed to organic pulp materials. By the close of 1989, the provinces of Ontario, British Columbia, and Quebec had announced both interim targets and final standards of 1.5 kg/ADt by 1993 or 1994 (Harrison 1996b).

While the precise loading weights were a matter for intense debate in both government and industry circles, all parties sought ultimately to achieve "non-detectable levels" of dioxins at point of discharge. As Ottawa developed its own draft regulations in 1990, it identified this goal with an equivalent of 2.5 kg/ADt. After several years' delay, the 1992 regulations designated dioxins a toxic substance under CEPA. However, the federal government stopped short of stipulating a federal AOX standard that would provide a floor level of regulation from coast to coast. Nonetheless, Ottawa's 1992 package, in requiring partial substitution of chlorine dioxide (ClO_2) for elemental chlorine (Cl_2), was considered equivalent to 2.5 kg/ADt. This choice was strongly influenced by industry and soft (conciliatory) province pressures and ran contrary to environmental groups and hard (rigorous) province positions.

The CPPA opposed a federal standard altogether and questioned the efficacy of the AOX parameter, arguing that it did not differentiate organochlorines of differing toxicities and contending that below the 2.5 level any organochlorines would essentially be inert. The industry lobby also joined with the provinces to ensure a "single window" enforcement of any parallel regulations. Certainly the lack of a designated federal AOX standard represented a political victory to producers in the soft provinces. The same applies to Ottawa's rejection of the more rigorous best available technology criteria, in favour of the reasonably available control technology. Harrison aptly sums up the political bargain, observing that "in the absence of provincial consensus, it was the least stringent rather than the most stringent provinces to which the federal government deferred" (Harrison 1996b, 494).

While mindful of regulatory politics and active in shaping statutory rules, the pulp and paper industry adopted its own business strategy for dealing with dioxins. In 1989, while the CPPA was pledging to meet any mandated government standards on dioxin emissions, member firms were applying their own transition measures, aimed particularly at reassuring customers in export markets. Even before the government's standard-setting exercise, industry made a voluntary commitment to eliminating detectable dioxin levels from mill effluents. This centred on the reduction or elimination of elemental chlorine and its replacement with chlorine dioxide or with oxygen delignification treatments. It was suggested in the US that elemental chlorine-free (ECF) bleaching could achieve an AOX level as low as .6 kg/ADt. In Canada, the Paprican research centre estimated that non-detectable dioxin levels could be achieved with as little as 40 per cent ClO_2 substitution. Over the decade, producers made considerable progress toward their announced goal, as the global proportion of ECF pulp output rose from 4 per cent to 64 per cent (Harrison 2002).

Environmental groups such as Greenpeace defined the dioxin issue in a very different way, choosing to focus on the entire family of organochlorines. Their solution called for an immediate shift to "totally chlorine free" (TCF) pulp-making processes (Greenpeace International 1998). This standard represented quite a different hurdle for producers, involving as it did a wholesale shift to peroxides and other bleaching agents together with enhanced oxygen delignification. By the end of the decade, some 6 per cent of pulp output was totally chlorine free. Today, the environmental movement looks beyond end-of-pipe solutions to effluent-free or closed-loop production. This can be achieved either by new greenfield design or by retrofitting current mills with "kidney"-style treatment processes.

THE SECOND GENERATION: THE PULP AND PAPER MILL
EFFLUENT REGULATIONS OF 1992

As part of Environment Canada's regulatory revival, an internal report had
been commissioned in 1986 into the impacts of the PPERs. Known as the
"Sinclair Report," it offered a comprehensive review of the past 15 years.
While the regime was pronounced an "administrative success," the study
frankly noted the limitations of compliance agreements and the loopholes in
the tiered regulatory template. In his conclusions to this report, which was
not intended for public release, the author recommended that a revised set
of regulations be made universally binding. The Sinclair Report entered the
public domain after a copy was leaked to Greenpeace, who released it with
much critical fanfare in the spring of 1989 (Macdonald 1991). The discovery
that most Canadian mills were not covered by the 1971 PPER regulations and
that progress toward compliance had slowed or halted did nothing to bolster
public confidence in the efficacy of regulation. Overnight, the pulp and paper
industry seemed exposed as a "black sheep" sector on pollution (Doern 1995).
The highly activist minister of the environment, Lucien Bouchard, responded
to the growing furore by promising both a dioxin standard in 1990 and a
revised set of compulsory PPERs by 1994. This unusual "burst of ministerial
leadership," as it was described, coincided with the buildup to Ottawa's 1990
Green Plan and Environment Canada's brief but energetic moment of policy
hegemony (Doern and Conway 1994).

If the second generation of pollution regulation was triggered by the
(re)discovery of new toxic substances and fuelled by the concern for more
precautionary research and vigorous enforcement, it also prompted a reas-
sessment of the earlier pulp and paper effluent regulations. The underlying
aims included the following: new binding rules applicable to all pulp and paper producers; revised discharge standards at higher values where appropriate; design of an environmental effects monitoring system on mill discharge
impacts to facilitate regulatory assessment; and provision for transitional
exemptions to allow added time for non-compliant plants to adapt.

On 7 May 1992, three sets of new pulp and paper regulations were issued.
SOR-92-267, or the Pulp and Paper Mill Effluent Chlorinated Dioxin and
Furan Regulations, set the non-detectable levels standard described in the section above. Accompanying this was SOR-92-268, or the Pulp and Paper Mill
Effluent Defoamer and Wood Chip Regulations, which prohibited certain
chemical treatments that would enhance the formation of organochlorine
compounds more generally. Each of these was issued under the authority of

CEPA. The third set of Pulp and Paper Mill Effluent Regulations (hereafter PPER-II), replaced the original 1971 regulations under the Fisheries Act. They set new levels for TSS and BOD discharges, though these remained tied to production rates, and also required secondary treatment facilities at *all* mill operations. While the shift to full binding coverage marked a major policy shift, it was a more subtle provision than was commonly recognized. Without legalized limits for pollutant discharge, the release of any effluent deleterious to fish would amount to a violation under s.36(3) of the Fisheries Act. From an opposing, critical, perspective, this was precisely the complaint made by the Sierra Legal Defence Fund in 2000. Among the lead Canadian ENGOs, closed-loop effluent control was now a consensus objective for pulp and paper. From this angle, the PPER-IIs were tantamount to "legalized pollution."

All three regulations took effect at the close of 1992. However, they also made allowance for case-by-case permissions to postpone compliance for one year and extraordinary authorization to postpone compliance for three years until the end of 1995. These options were invoked by more than half of all Canadian mills, with 91 winning the one-year extension and 79 of these later gaining the three-year authorization as well.

Of considerable significance was the environmental effects monitoring plan. It aimed to institute an ongoing test of the impact of effluent on fish and fish habitat in order to light on the need for added controls in the future and to permit the first assessment in terms of receiving water quality. Since the design details were not yet complete at the time of issuance, this was left for Environment Canada to settle after 1992.

Any assessment of PPER-II needs to begin with the fact that, whatever their content, the 1992 regulations had the effect of shielding the industry from prosecution under s.36(3) of the Fisheries Act that made it illegal to deposit substances deleterious to fish. As the auditor general observed,

> Without the new regulations, 90 per cent of the pulp and paper mills could have been subject to a potential threat of prosecution. Departmental officials told us that "there have been very few convictions under the Fisheries Act." Nevertheless, implementing the regulations afforded mills who met the regulatory requirements with some protection, by allowing the discharge of prescribed amounts under specified conditions. (Canada 1993, 635)

However much the industry might resent the intrusive nature of these regulations, and the costs of compliance, they were far preferable to the "zero discharge" provision that the Fisheries Act contemplated. And while grandfather provisions were no longer viable, the seemingly infinite flexibility of legal deadlines meant that a process announced in 1989 and negotiated for almost three years gained a similar period for phase-in, granting the pulp industry a full six years, or more than a typical business cycle, to put its effluent house in order.

Not surprisingly, the producer and environmental perspectives continue to be diametrically opposed. The former contends that almost $6b was invested in pollution abatement systems from 1989 to 1997, principally in the installation of secondary treatment facilities. Over this time, BOD loadings were reduced by 94 per cent while TSS loadings fell by 70 per cent, AOX by 88 per cent, and water use by 48 per cent. In addition, dioxin and furan releases were non-detectable by 1995 (FPAC 2001).

On the other hand, the ENGO constituency involved in pulp issues focussed on the central fact that significant levels of pollution discharge continued to occur. The Sierra Legal Defence Fund report, *Pulping the Law*, obtained confidential data by utilizing access to information procedures that established that thousands of violations of discharge standards took place in the 1990s. While 32 mills were responsible for more than 2,000 violations, only seven mills were charged. The group concluded that "Ottawa has all but abandoned its efforts to enforce [the 1992] laws" (Sierra Legal Defence Fund 2000, 4). This point has been repeated, at regular intervals, in subsequent years.

RETREAT FROM REGULATION:
FLEXIBLE COMPLIANCE WITH VOLUNTARY STANDARDS?

Canadian national policies for the control of pulp and paper pollution have been strikingly cyclical, swinging from periods of aggressive regulation-making pushed by surging public concern (1967-71; 1989-94) to periods of inertia when the combination of industry and provincial government resistance sapped Ottawa's will to extend and enforce its mandate (1975-85). Recently it has been suggested that pollution control policies in general have entered a new political phase. Harrison has described a "trend away from regulation" that accelerated from the mid-1990s. In addition to the pressures already noted, this was furthered by several state-centred programs including the Chrétien-Martin program review, which squeezed Environment Canada budgets; the Canada Wide Accord on Environmental Harmonization, which ceded the lead implementation role for standard-setting and enforcement to the provinces; and the rise of a new guiding policy philosophy in Environment Canada. Where earlier paradigms assumed an inherently adversarial relationship in which direct legal controls were essential to securing corporate compliance, this new view stressed the market-driven advantages of sophisticated environmental performance and the need for Environment Canada to play a facilitating as opposed to a coercive role in supporting this transition (Harrison 1999). In practical terms, this entails more flexible paths to regulatory compliance, such as negotiated plans under CEPA; the possible introduction of market-based instruments, such as discharge fees, user fees, and tradeable permits; reliance on lower threshold regulations, such as corporate information disclosure to national registers and firm-based pollution prevention plans; and voluntary business initiatives in self-regulating or externally audited settings. Such arrangements include eco-labelling schemes, third-party certified environmental and sustainable management schemes, and multi-stakeholder programs like the Accelerated Reduction/Elimination of Toxics (VanNijnatten 1998).

All of these have contributed to the contemporary debate over the parameters of pollution protection policy and the mix that best serves the public interest. Pulp and paper industry interests now argue vigorously in favour of these "new" as distinct from the "old" approaches.

Clearly, every newly authorized regulatory or program measure joins an already sizeable body of pollution control policy for the pulp and paper industry. The result can be viewed as a layered regime, an uneven and potentially contradictory assemblage that can be staffed, financed, and enforced in varying degrees. The value of an industry or sectoral analysis can be readily seen when the pulp and paper mill effluent regime is considered over time. A number of signature characteristics distinguish PPER-I from PPER-II. Bolstered by compulsory application and the environmental effects monitoring system to incorporate receiving water impact data, the 1992 regulations were a significant step forward. Certainly there remains the question of enforcement and compliance techniques, yet, in the face of the significant improvements in discharge compositions during the 1990s, it can hardly be dismissed as an abdication.

Of equal political significance is the fact that PPER-II was in place before the current metapolicy framework was fully articulated. While the CPPA (now known as the Forest Products Association of Canada) will continue to press for more flexible, non-coercive measures, it is difficult to conceive a political scenario enabling the PPER edifice to be dismantled wholesale. So long as Sierra Legal Defence Fund (2000) can effectively present a spectrum of corporate performance ratings, ranging from the best—Millar Western at Meadow Lake—to the worst—Tembec at Temiscaming—the role of a federal regulatory mandate will remain a matter of pressing political debate.

The Pulp and Paper Modernization Program, 1979-1984

During the late 1970s, Canada's national and provincial governments collaborated on a major initiative to promote the modernization of existing pulp and paper mills. Over the life of the program, more than half a billion dollars of public funds were spent in efforts to lever a far larger private capital investment in new production, pollution treatment, and energy conservation facilities. Although sectoral modernization programs became more common over time, this was the first targeted manufacturing sector initiative to emerge from Ottawa's industrial policy deliberations. In pulp and paper, it was a product of sustained consultation between business interests and state authorities, and it attracted a considerable level of federal-provincial support. As such, it reveals much about how these key structures interact on an issue of high political priority. In retrospect, the pulp and paper modernization program (PPMP) has been criticized for impeding a necessary rationalization of plant and firm units at a critical historical moment. At the same time, it is a revealing example of the dynamics of a broad industry-wide public policy initiative.

While the process of policy formulation will be detailed below, it is important to appreciate the broader economic and political context that made the PPMP possible. The 1970s were a turbulent decade for this industry, which had enjoyed a prolonged period of postwar growth. In addition to the stagflationary pressures that defined the macroeconomic climate, pulp and paper faced the threat of soaring energy costs after the 1973 OPEC crisis and chronic uncertainty from the volatility of currency exchange rates in the period which followed the floating of the US dollar. For Canadian producers, particularly in the central and eastern regions, these difficulties were exacerbated by the dramatic expansion of manufacturing capacity in the southeastern US states, where fast-growing pine forests provided the fibre stock for modern high-volume mills, which also enjoyed added transportation and wage advantages. Thus, when Canadian export sales began to decline, there were concerns about a structural as well as a cyclical downturn. While Ottawa worried about the health of a huge manufacturing export industry and earner of foreign exchange, the leading host provinces saw dozens of single industry towns in jeopardy should pulp and paper decline. They turned to the industry for advice.

As we have seen above, the forest industries have a long history of political organization and have pursued a wide variety of contacts with the state. On the forest management and harvesting side, the major contacts have been provincial natural resource ministries together with the federal forestry service, which functioned within the DOE from 1972 to 1993. By contrast, the policy jurisdiction over forest manufacturing has fallen principally to provincial ministries of trade or industry, while in Ottawa it was shared by the Department of Trade and Commerce and the Department of Industry, which were merged from 1969-82. It was on the latter side that the modernization problem was grasped. A critical event was the creation, in 1974, of the Forest Industries Development Committee, which brought together officials from eight federal departments and the provinces. This committee forged much of the basis for shared understanding and collaboration among governments. Moreover, since it was also charged with consulting private-sector interests, it emerged as a central point of contact between business and state authorities (Sidor 1981). As well, both Ontario and Quebec had set their own industry-government task forces in motion by 1978, a fact which strengthened their capacities to act.

Before any collaborative state action was possible, however, it was necessary to fashion a shared appreciation of both the problems and solutions. In this, the work of the Forest Industries Development Committee was crucial. Through repeated meetings and studies, an increasingly coherent interpretation and definition of the issue emerged to guide policy formulation once the path was cleared for action. This "model," if it may be so described, identified the key problem for Canadian pulp and paper producers as a lack of access to the capital necessary for modernizing their plants. The combined impact of reduced profits, more attractive (offshore) investment sites, and a decline in readily available timber was blocking firms from renewing their capital assets in Canada (Canada 1978a; Ontario 1978).

At this point, it is important to appreciate the intersection of pulp and paper policy with Canada's broader initiative for industrial policy. Throughout the 1970s, there was a general concern with the decline of the country's manufacturing sector. Variously defined as a problem of foreign economic ownership or de-industrialization, it became a major policy preoccupation in Ottawa, if not always in the provinces. Though not without its market-oriented critics, both in intellectual and government circles, the desire for a coordinated mix of policy measures framed to overcome structural barriers to industrial growth was embraced by key cabinet and central agency figures by 1972. It is possible to chart the rise and fall of this initiative from the grand ambitions for a comprehensive industrial blueprint to the far more modest and pragmatic search for limited sectoral enhancement (French 1980). It was driven by a growing appreciation of the institutional obstacles, ranging from competing policy priorities, such as wage and price controls and other fiscal restraint measures, to federal jurisdictional boundaries, to bureaucratic antagonisms in Ottawa. Indeed, an early proposal for a pulp and paper modernization scheme, brought to the Trudeau cabinet in 1973 by the minister of industry, trade and commerce, was defeated amidst considerable bureaucratic opposition from the Department of Regional Economic Expansion or DREE.

By 1978 the political circumstances had changed. The industry department had launched a new, bottom-up initiative, based on federal, provincial, and industry-level consultations through 23 sectoral task forces, including one for the forest industries. In Ottawa, this was known as the "Tier One/ Tier Two" process. Each Tier One committee was guided by a sector analysis prepared by industry department staff, though the committee members were responsible for fashioning consensual policy recommendations. Naturally the results varied considerably from one sector to another. The forest industry task force was able to draw on a tradition of prior consultation. Perhaps not surprisingly, its members fashioned a set of policy incentives to attract modernization capital. Noting the mounting cost disadvantages that afflicted Canadian producers, particularly in export markets, they identified a series of public policy adjustments that could promote improved performance (Canada 1978b). Its centrepiece was a revised tax regime to encourage capital investment and industrial research and development. However, important proposals were also directed at streamlining pollution regulations, greater state support for intensive silviculture, and the adjustment of Ottawa's new trade and competition policy initiatives in light of the modernization imperative. This increasingly consensual policy diagnosis was later challenged as empirically inaccurate (de Silva 1988); however, it was no less influential in its time for this fact. Indeed, once the political will existed for policy intervention, many of the intellectual and institutional obstacles to action had already been cleared away. Responding the following year, the federal government accepted the diagnosis, but favoured a capital grant program as opposed to tax relief (Canada 1979).

Even before this formal response had been released, considerable intergovernmental consultation took place. The Trudeau cabinet directed DREE, as Ottawa's prime agent for collaborative initiatives with the provinces, to generate a proposal. Anticipating and perhaps hoping to force an early federal announcement, Quebec declared its own modernization plan in June 1978, involving some $450m in grants and loan guarantees. Quebec also approached DREE to indicate its interest in a new subsidiary agreement to the Canada-Quebec General Development Agreement. Ontario also announced a plan of its own in January 1979.

In response, Ottawa decided to support a national program for pulp and paper modernization. The following month Robert Andras, a member of

Parliament from the pulp and paper dependent region of Northern Ontario and the federal minister in charge of the new cabinet Board of Economic Development Ministers, announced a $235m federal contribution to a collaborative PPMP. Once provincial contributions were taken into account, the total public commitment was almost $550m, which was to be distributed as grants to trigger far greater capital commitments by the pulp and paper firms. The federal position was coordinated by the board, while the program was delivered by a new set of subsidiary agreements to the General Development Agreements already negotiated between DREE and each provincial government. The first of these, with Ontario and Quebec, were finalized in record speed by May 1979, while New Brunswick followed in 1980 and Newfoundland and Nova Scotia in 1981. It was in Quebec and Ontario that firms were most bluntly weighing mill closures, and those provincial governments were anxious to act while the rising business cycle offered investable surpluses.

The actual terms of the program, depicted in Table 2-2, were revealing. From the outset, Ottawa insisted that this program was not intended to alter regional shares of pulp and paper production. In fact, the federal contribution was partitioned initially into four distinct allotments for the Atlantic ($60m), Quebec ($90m), Ontario ($46m), and western ($39m) regions. Once it became evident that the western provinces and western industry were less vulnerable and less interested in the program, Ottawa offered them alternative forest development schemes, while refocusing the PPMP on the eastern market. After agreements were concluded with Newfoundland and Nova Scotia in 1981, Ottawa added an additional $45m to the Atlantic allotment, raising the total federal commitment to $290m.

Ottawa was equally concerned that the grants serve to improve productivity rather than simply add extra capacity to existing plants. For this reason, newsprint expansion was excluded specifically from the list of eligible expenditures. In fact, this proved to be a difficult distinction to apply, and modest increases in capacity were countenanced when they figured as a secondary result of the installations. Although the main goal of the PPMP was to promote the installation of modern machinery and processes, investment in pollution abatement and energy conservation technologies were also eligible. Ultimately, however, almost 80 per cent of the funds supported modernization expenditures, with approximately 20 per cent directed mainly to pollution abatement measures. For any project, public support was limited to 25 per cent of total planned expenditures.

As Table 2-2 reveals, the cost-sharing arrangements varied by province. Since Quebec and Ontario had committed funds of their own in advance of

Table 2-2

Allocation of Funds Under the PPMP, 1979-1985

Province	Fed/Prov Share	Federal ($)	Prov ($)	Total ($)
Newfoundland	90/10	38.265 m	4.252 m	42.517 m
Nova Scotia	80/20	14.992 m	3.748 m	18.740 m
New Brunswick	80/20	42.600 m	10.650 m	53.250 m
Quebec	56/44	135.085 m	106.143 m	241.228 m
Ontario	33/66	62.163 m	124.326 m	186.489 m
Total		293.105 m	249.119 m	542.224 m

Source: de Silva 1988, 105.

Ottawa's decision, they ended up absorbing a larger proportionate share, which in the case of Ontario actually amounted to two-thirds of the total. The Atlantic agreements reflected the more traditional federal predominance in economic programs. Anxious to stimulate investment while the 1977-79 recovery lasted, Ontario also pressed for substantial "up-front" payments and ultimately designated its contributions to this end, while the federal shares were dispensed over three years after the corporate commitments had been made.

One of the most important conditions for grant eligibility was the purchase of machinery made in Canada. This reflected one of the key attributes of industrial policy: that linkages be forged among distinct chains of commerce. It may also explain the overriding enthusiasm in Quebec and Ontario, where these capital goods suppliers were concentrated. Another key dimension was the choice of cash grants, rather than tax aids, to supply the incentive, despite the persistent preference of private-sector interests for the tax route. A number of important considerations had a bearing on this key decision. One sprang from the policy inclinations of competing bureaucracies. The federal finance department argued successfully that Ottawa had already enriched the tax relief available to forest companies. Provisions in its 1978 budget extended the investment tax credit and research and development tax credit as well as the two-year tax deduction for pollution abatement equipment. The key operational departments, regional economic expansion and industry, were more familiar with grants as policy instruments that could be closely tapered to business behaviour. Moreover, grant programs could be readily designed to achieve joint federal and provincial participation for maximum impact. Another important factor was the speed with which the program could be launched and modernization begun. Tax changes were locked into an annual budget cycle and could not offer financial relief for another year after that when firms claimed the relief, whereas grant programs could be launched quickly, with cheques written as soon as the modernization expenditures were made. The PPMP is a rather extreme example of this, since it was launched in Ottawa without any new authorizing legislation (Canada 1986). A third consideration was the question of precision targeting. Tax relief is valuable to profitable firms with a net income to shelter but offers no help to firms reporting an annual loss. Since a grant injects cash into any eligible corporation, it has a wider potential field of application, including many of the marginal firms in greatest need of modernization. Moreover, the grant applicant must enter into detailed negotiations prior to approval, enabling state authorities to taper their influence to particular firms. For instance, the province of Ontario actually refused assistance to the Reed Corporation on the basis of its poor pollution record and eventually persuaded Reed to sell off its northern Ontario mill.

ltimately the PPMP experience raises a number of intriguing questions.
⸱⸱ibute to the desired goals? Perhaps the main concern was to sta-
ective modernizations, a set of aging pulp and paper mills
competitive position raised the prospect of eventual closure.
. the very least, closure was postponed; that is, firms opted to
of those mills by an additional decade or more. But was the
,ary to achieve this outcome? At least one study has argued that
zation grants failed in their aim of incrementality (i.e., of levering
estment through public incentives), since only firms with prior
ents to modernization took the plunge (de Silva 1988). Even if this
,d, there may have been less transparent policy reasons for the PPMP.
.   .mple, the new pollution standards emerging in the 1970s, discussed in
the previous policy case, placed a burden on Canadian forest firms, which had been virtually unregulated in the past. Was the PPMP designed to reassure the pulp and paper sector of its place on the policy agenda? Furthermore, not all firms availed themselves of the modernization grants; in fact, many of the oldest mills were left untouched. While this may be construed as evidence of failure (de Silva 1988), it is equally possible that the PPMP had an informal screening effect, touching the most competitive operators while by-passing the truly ancient. The terms of the program conferred extensive administrative discretion on senior government officials (Webb 1990). There is also the factor of differential corporate strategies, which certainly deserve closer consideration, since the response of a captive newsprint or market pulp mill producer with guaranteed outlets for its output and having fully recovered its

sunk capital may well contrast with the response of a commodity producer having to win contracts in a competitive market and confronting off-shore rivals enjoying cost advantages. It is notable that such questions have continued to plague industrial policy initiatives in the energy, aerospace, automobile, and telecommunications sectors.

From the outset, the PPMP was intended to have a five-year duration. However, its termination was in a sense over-determined by wider currents, both in forest industry politics and in Ottawa's international trade strategy. For some time the leading elements in the forest industry had expressed concern about the use of investment subsidies for plant modernization. The Forest Sector Advisory Council, as the recognized clientele of the departments of industry and forestry, had called for the discontinuation of capital improvement grants and the adoption of tax-based incentives in their place. As the Council put it in 1985, "the provisions of direct discriminatory grants and subsidies to individual companies for capital investment for both new facilities and for modernization should be ended. They create inequities, result in the postponement of private restructuring decisions, reinforce this sector's tendency toward oversupply and reduce the overall investment attractiveness of the sector" (FSAC 1986, 14). This coincided with Ottawa's growing awareness of the potential for foreign trade disputes, based on the market-distorting effects of targeted business subsidies. (This is discussed for steel in the chapter to follow.) For the forest sector, it was brought forcefully home in 1986, when large US lumber producers successfully challenged Canadian softwood lumber imports on the basis of stumpage-related subsidies to trade. It was a short step from lumber to pulp and paper and from stumpage to modernization grants. Put simply, Ottawa was unwilling to further risk the export prospects of its leading industrial export engine. In 1987 the federal government terminated such grants as part of its new Forest Industry Policy (Canada 1997a)

Conclusion

After reviewing the historical strength of pulp and paper production structures in Canada, together with the durability of leading firms in adapting to changing fortunes, it might be wondered whether this sector has ever had difficulty in realizing its political interests. Certainly the economic force of the industry — as a leading manufacturing producer, a leading rural employer, and a leading exporter and foreign exchange earner — has opened doors at the state level and ensured that corporate concerns were well represented. It would be too much, however, to suggest that pulp and paper producers have enjoyed unquestioned dominance or hegemony throughout the modern period. One need only reflect on the traditional sources of political challenge to a core industry to sense the many possible complications: internal divisions within the core producer bloc; other-industry challenges at home; non-business rivals such as labour, consumer, and environmental interest groups; or same-sector rivals from abroad. At least three of these play a role in the policy cases just reviewed.

With pulp and paper, it could be argued that the industry has largely won its battles with the federal and provincial governments, particularly when the policy questions involve business arrangements *within* the sector. Wyn Grant offered an apt description of the business-government relationship for the forest industries as one that combines "a high level of organizational development of business interests with a relatively fragmented state authority" (Grant 1990, 120). However, the result has been less decisive in the political contests with the public interest groups and environmental social movements that have achieved growing stature over the past generation. This underlines the contrast between the "old" forest politics, in which close industry-state clientele relations prevailed, and the "new" politics, in which a more diverse and complex policy network prevails.

The pollution issue offers a fascinating example of a policy network evolving over more than 30 years. It began as a case of classical clientelism. Pulp and paper capital entered the debate from an established position of economic strength, with representational allies in both federal and provincial departments. During the opening phase, the paper industry pursued a variety of tactics: playing the two governments off against each other, capitalizing on a virtual monopoly of technical expertise, and securing the grandfathered status for existing mills that ended up lasting for decades.

Then, however, the pulp and paper pollution issue was redefined. The old narrative of benign discharge—"the solution to pollution is dilution"—was abandoned in the face of increasingly threatening evidence of toxicities. Moreover, environmental protection advocates played the crucial role in this policy redefinition, joining the debate from both the scientific (Carson 1962, among others) and advocacy (nature and wilderness protection) bases. The new narrative involved government regulation to protect the public interest, and a new set of state agencies assumed the lead role. Nevertheless, as the policy cycle shifted from agenda setting to design and implementation of the PPERs, clientelism was not entirely displaced.

By the second generation of regulatory politics, the environmental community had broadened and deepened. New voices emerged, and improved communication networks and capabilities for rapid political response extended their reach. On pulp and paper, the lead ENGOs included Greenpeace, the Sierra Club of Canada, and the Sierra Legal Defence Fund, but coalition campaigns allowed political action on an unprecedented scale. Non-business interests had achieved a permanent institutional presence, enjoying ready access to the electronic media and building an impressive base of volunteer members. The ENGOs were, however, reluctant to shift from a political stance grounded in pressure pluralism to a place at the clientele table.

This continued into the third era, when certified third-party standard-setting emerged as an alternative regulatory regime. Environmental authorities embraced these arrangements at a time when government's capacities to staff and deliver traditional regulatory systems have been questioned. This is emerging as a major problem in environmental governance. It remains to be seen, however, whether this represents a new setting for the reassertion of clientelistic networks and how the environmental advocacy community will respond.

So long as the leading productive sectors remain dynamic and competitive, their core business practices tend to remain off the policy radar. This enables collaborative concerns to predominate on secondary issues where clientelistic accommodation is routine. However, any signs of structural faltering or stress opens the possibility of new policy agendas that cannot easily be accommodated within traditional channels. In our second case, the focus shifts to the policy implications of the perceived competitive decline of Canadian pulp and papermaking facilities, also beginning in the 1970s. Coinciding with Ottawa's passing engagement with the politics of industrial strategy, we see a rare attempt by state authorities to reshape some basic parameters of a core manufacturing sector. This proved more episodic and less durable than the pollution problem, though it posed some intriguing questions about the role of government in orchestrating sectoral change.

This limited set of policy cases suggests that the prevailing network for pulp and paper should be conceptualized on at least two levels: business clientelism and populist pressure advocacy. Neither offers a sufficient characterization for all issues on its own. The two-tier structure may operate in an unsteady equilibrium, enabling issues to be grasped on either scale.

Key Terms and Relationships

Accelerated Reduction/Elimination of Toxics: A voluntary program linking the federal DOE and several industries, including pulp and paper, beginning in 1994 as an experiment in non-regulative reform.

Canadian Environmental Protection Act (CEPA): The second generation federal pollution protection statute enacted in 1988, which tackled toxic pollutants and extended the scope of regulation.

Environmental Effects Monitoring: A management tool introduced by CEPA, which measures pollution protection techniques against local aquatic ecosystem health.

Harmonization Initiatives of the Canadian Conference of Ministers of the Environment: Began in the late 1980s with sectoral rule-making on issues such as organochlorines and expanded with the Canada-Wide Accord of 1998.

Income Tax Capital Cost Provisions (federal): To encourage the adoption of pollution prevention technologies by allowing their accelerated write-off in as little as two years.

Organochlorines: Dioxins and furans, compounds related to chlorine bleaching, were identified as highly toxic in 1987.

Pulp and Paper Effluent Regulations (PPER): The principal instruments for regulating mill effluents since 1971, setting standards for suspended solids, biochemical oxygen demand, and acute lethality to fish.

Pulp and Paper Modernization Program (PPMP): Developed in the late 1970s as a federal-provincial incentive program for technology upgrading in the mills of eastern Canada, at a time when their competitiveness was coming under question.

Recycling regulations for paper products: Began to appear in the late 1980s in both Europe and North America. Bolstered by consumer preference, they set minimal standards for recycled fibre use in pulp and papermaking.

Stumpage charges on crown forests: Charges paid by logging operators to provincial state forest owners.

Tariffs: From the elimination of the US newsprint tariff (1913) to the perpetuation of significant Canadian tariffs on fine papers, paperboard, and tissues, these have played a leading role in shaping the structure of the Canadian industry.

CHAPTER THREE

Steel Politics

The Industry at the Millennium

For the steel industry it might be said that the more things change, the more they stay the same. This was the paradox of steel politics at the close of the twentieth century after the massive structural changes that had taken place in the decades following 1970. The open hearth furnaces that had dominated big steel production for more than a century were phased out in North America. The mini-mill (electric arc) steelmakers, which first emerged on the lower-value commodity fringe of the industry, rose to virtual parity with the integrated US producers, invaded the high-end sectors of flat-rolled steel, and earned leading-edge profits while doing so. The Canada-US Free Trade Agreement (FTA) of 1989 and the North American Free Trade Agreement (NAFTA) of 1993 created a continental market for industrial goods like steel, with the promise of a level playing field for cross-border trade. Producers even began to join into bigger corporate blocs through a series of massive European and Asian merger deals beginning in the late 1990s.

Yet, despite these striking developments, the overall image, not to mention the political problems, of world steel remained much the same. To describe it as a less than efficient sector burdened with excess capacity, miserable financial fundamentals, and defensive political proclivities, was not inaccurate. In the North American market, steel prices appeared to be disconnected from the general business cycle, and volatility was the watchword. The cost structures of the big integrated mills were significantly higher than the most efficient world steelmakers, so even when demand was strong profitability was problematic. In 2001 Bethlehem Steel, the traditional number two US firm, had a negative net worth of $170 million. A year later, it followed other top US integrated producers, like LTV and National Steel, into court-supervised bankruptcy protection. After 1997, the level of overseas shipments into the US (and Canadian) markets grew quickly, and prices fell in response (see Figure 3-2).

In the US, steel was a tale of two sectors: the travails of integrated steelmakers and the triumphs of mini-mills. This could be seen both in business and politics. In the market, the leading mini-mill operator, Nucor, challenged integrated leader US Steel for top place among US producers. On the government relations front, the divergence of policy agendas intensified. The integrated firms, represented by the American Iron and Steel Institute, wanted Washington to rely principally upon the defensive trade policy tools of anti-dumping and countervail, along with special import protection measures under section 201 of the US Trade Act of 1974. Many also relied upon the courts to hold off creditors under the US's uniquely forgiving Chapter 11 bankruptcy provisions, in the hope that market conditions would improve while new business plans were imposed. However, these were inherently short-term measures that failed to address the deeper structural malaise.

The mini-mill producers, represented by the Steel Manufacturers Association, called for global negotiations on capacity reduction in order to phase out inefficient producers. From this perspective, the strength of offshore imports was a symptom of commercial weakness more than its cause. These innovative, low-cost operators had little patience for the traditionalists in big steel who, they felt, were feeling the results of poor strategic and managerial choices. Nor did the association agree with calls for massive government loan guarantees or subsidies to defray the crushing burden of retirement "legacy costs" that big steel faced from earlier eras.

The election of George Bush Jr. to the US presidency was a signal event. Nominally, he was an open market, free trade Republican. Yet he also had political debts arising from pledges to steel communities in key electoral college states like Ohio and Pennsylvania and lesser states like West Virginia and Indiana. In June 2001 Bush paid back these debts by launching an investigation into the role of steel imports as a substantial cause of serious injury to domestic firms. Nine months later, in March 2002, he went beyond the International Trade Commission (ITC) recommendations, ordering tariffs of up to 30 per cent on an array of landed steel products. While this was described at the time as a bold stroke of relief for the domestic industry, it was really a lowest common denominator choice. It by-passed the more complicated strategic option, which was being advanced in many quarters, of reorganizing the structures of integrated steel production through mergers, closures, realignments, and negotiated reductions in capacity. Instead, Bush provided short-term relief by blocking out steel imports in the hope of releasing market share for ail-

Table 3-1

Steel Producers: World (2002) and Canada (2000)

Rank	World Producers (2002)	2002 crude output
1	Arcelor (Fra-Lux-Spain)	44.0 m tonnes
2	LNM Group – Ispat International (worldwide)	34.8 m
3	Nippon Steel (Jap)	29.8 m
4	Pohang Iron and Steel Co. (So. Korea)	28.1 m
5	Shanghai Baosteel (PR China)	19.5 m
6	Corus Group (UK)	16.8 m
7	Thyssen Krupp (Ger)	16.4 m
8	NKK Corp (Jap)[a]	15.2 m
9	Riva Group (Ita)	15.0 m
10	US Steel (US)	14.4 m
Canadian Producers (2000 Capacity)		
46	1. Steel Company of Canada – Stelco Inc.	4.6 m tonnes
51	2. Dofasco Inc.	3.6 m
	3. IPSCO Inc.	3.5 m
	4. Co-Steel Inc.	3.5 m
	5. Algoma Steel Inc.	2.55 m
	6. Ispat Sidbec Inc.	1.6 m
	7. QIT – Fer et Titane Inc.	1.1 m
	8. Slater Steels Inc.	915,000
	9. Ivaco Inc.	900,000
	10. Atlas Tube Inc.	680,000
	11. Gerdau (Courtice / MRM) Steel Inc.	640,000
	12. LTV Copperweld	550,000

[a] In 2003, NKK completed a merger with Kawasaki, resulting in JFE Steel Corp., with a combined 26 m tonnes.

Sources: International Iron and Steel Institute 2003 (World); Cooke 2001 (Canada).

ing home producers. A new cycle of instability was launched. Whereas the abandonment of LTV would have removed millions of tonnes of inefficient steel production, the company was now sold to new owners and prepared for relaunch. Where industry leader US Steel had been exploring a mega-merger with weakened rivals Bethlehem, National, and Wheeling-Pittsburgh, it broke off talks and returned to the competitive fray.

In Canada, the 1990s closed on a more benevolent, though still worrying, note. Here the competitive gap between integrated and mini-mill operations was less stark, and the industry shared a common voice in the Canadian Steel Producers Association (CSPA). Over the course of the decade, annual output rose an average of 8 per cent compared to 4 per cent in the US. Apparent consumption (domestic output plus imports minus exports) doubled to 20m tonnes. In 2001, Canada's crude steel output fell just short of the European number four nation Spain and ahead of number five Britain. Canada trailed just behind Asian number five Taiwan and finished ahead of Mexico. Almost all leading Canadian firms (listed in Table 3-1) shared in this advance. The most dramatic change saw mini-mill producer Interprovincial Steel Company (IPSCO) more than triple its output and become multinational by opening two major greenfield facilities in Iowa and Alabama. By the year 2000, both IPSCO and Co-Steel had overtaken integrated steelmaker Algoma on the rated capacity ladder. Meanwhile, the two industry leaders—the Steel Company of Canada (Stelco) and the Dominion Foundry and Steel Company (Dofasco)—invested heavily in capital upgrades, increasing their shares of coated sheet steel for the automobile sector and promoting new products in the construction field.

Despite such strong production gains, the import problem intensified as the foreign steel share grew from an average of 25 per cent of total sales in the first half-decade to 39 per cent in the second. The president of Stelco, James Alfano, spoke of a "steel crisis [that] has gripped the United States and Canada since 1997" (Alfano 2002). In response, Canadian producers regularly petitioned the Canadian International Trade Tribunal (CITT) for relief from dumped steel imports. In the period from 1997 to 2002, there were ten anti-dumping cases, seven of which brought duties averaging 31 per cent on shipments from 28 countries.

From an industry point of view, however, the overall results fell far short of satisfaction. Both company leaders and the CSPA criticized the CITT for its timidity in asserting the mandate to guard against unfair trade practices. For steelmakers and steel unions, any shipments that violated international rules represented lost sales and therefore inflicted economic injury. Yet for steel-consuming industries and import distributors, such shipments represented bargain deals on supplies that domestic mills were unlikely or unable to supply. Located at the epicentre of this complicated political vortex, the CITT faced constant challenges to its procedures and decisions at all stages of the steel product chain. On another front, Canadian steelmakers had to manage a complicated trade relationship with Washington. In 2001, they lobbied intensely for an exemption at the preliminary stage of Bush's tariff initiative, and, failing that, they pushed for exemption at the later stages where injury and penalty were determined. Back in Ottawa, the steelmakers pressed the CITT for stepped-up monitoring and Canadian safeguard action against the diversion to Canada of offshore shipments that were being shut out of the US.

Both the Canadian and US experiences point to larger issues concerning the future of global steel. Several of the world's leading *regional* steel markets have shown dramatic signs of corporate concentration. In Europe, three giant firms have been created lately by merger and acquisition: Thyssen Krupp in Germany in 1997; the Corus Group in Britain and Holland in 1999; and Arcelor in France, Luxembourg, and Spain in 2001. In Japan, the number two and three firms, NKK and Kawasaki, completed a merger in 2003 that made JFE Steel Corporation another global giant. In the US, Nucor took over several mini-mills, the largest being Birmingham Steel. Only in the integrated US steel sector has this trend been absent so far. The ongoing problem of legacy costs (the continuing obligations to laid-off and retired workers) loomed large in discouraging buyers. By contrast, the *world* trend toward inter-regional alliances is far less developed. Japan's NKK took over National Steel (US), Ispat International purchased Inland Steel, and US Steel purchased the Kosovice Works in Slovakia. Corus first agreed and then terminated a merger plan with SNC of Brazil. Nevertheless, some scenarios now sketch a future world of giant producers, no more than six or eight in all, with a couple active on each major steel continent (*CRU Monitor—Steel* 2002a).

Finally, the millennium has had to recognize the China factor. The ongoing liberalization of the mainland Chinese economy after 1978 created an explosive demand for capital goods products and therefore for steel. In the period since, the growth rate has outpaced all others in Asia. Despite a surge of domestic steel output, which jumped 25 per cent between 2001 and 2002, China's large net deficit on the steel trade account has provided a crucial outlet for European and other Asian surpluses. The prospect of China consuming one-quarter of global steel output is not far away (*CRU Monitor—Steel* 2002b).

Overview

For a number of reasons, the steel industry is closely associated with the birth of industrial capitalism. It was an indispensable foundation of the manufacturing revolution in the latter half of the nineteenth century. Steel followed iron as one of the basic materials for modern manufacturing, given its strength, diversity of applications, and potential for mass production. Advances in metallurgy revolutionized the steelmaking process several times over, with profound effects on capital-labour relations, corporate profitability, and the geopolitics of production.

To chart the rise and subsequent decline of big steel is to map the history of industrialism, in spatial, trade, and sectoral terms. Britain's initial predominance gave way by century's end to rival commercial powers. The key names in the era of big steel were Carnegie and Gary in the US and Krupp and Thyssen in Germany. Later, in the aftermath of World War II, the US enjoyed pre-eminence, as heavy industry lay in ruins in Germany, France, Italy, Japan, the Soviet Union, and much of Britain. However, steel's recovery was a precondition for economic reconstruction both in Europe and the Pacific. In less than a generation, modern steel sectors with ever-greater capacity again anchored the leading capitalist and communist economies.

In the 1970s steel production reached an historic peak in the West. By then, it seemed to have entered the mature phase of the product cycle (Tiffany 1988). In the decades since, the world steel industry has revealed several contradictory trends. There have been massive struggles with surplus capacity in the Western capitalist states, a mounting flow of dumped commodity exports from the post-communist bloc, and dramatic growth in efficient new manufacturing capacity in Asia, beginning with Japan and South Korea and later extending to Taiwan and China. This coincides with a burgeoning international trade in steel products, together with a growing preoccupation, particularly in North America, with defensive trade measures such as anti-dumping and countervail (Hogan 1983). There has also been wholesale retrenchment among the formerly giant integrated producers, alongside explosive expansion by electric mini-mill producers. All of these factors shape the contemporary politics of steel, and they are examined in detail in the sections below.

Despite a growing global trade, the steel industry remains largely a nation-based form of enterprise. This owes much to the rather close relationship between state authorities and steel capitalists. More than most industries, integrated steelmaking has been imbued with complex political symbolism. At one level, the strength of heavy industry has been seen as a measure of economic might and sovereignty. In another sense, all states that have sought to emulate the British industrial model have tended to view iron and steel as a prerequisite. Consequently, any national capitalist growth strategy, whether in Europe, Latin America, Quebec, or Saskatchewan, must reckon with this sector. Tariff protection has played an important role in nurturing embryonic home markets, and state backing for "national champion" firms has taken a variety of forms. The postwar period also brought extensive state ownership of steel enterprise, particularly in western Europe and the developing world but also in some

Canadian provinces. However, the neo-liberal economic strategies of recent decades have reversed this, in a broad wave of steel privatizations.

Equally, the steel industry has been closely bound up in the development of the working class and the trade union movement. The organization of steel workers was pivotal in many nations, not the least in Canada, where modern industrial unionism began with campaigns in the steel and auto sectors and where union politics continues to shape both business performance and sectoral political initiatives. In fact, one important variable to be considered in exploring differences in US and Canadian industry performance is the role of organized labour.

While the US steel industry was unquestionably under threat, Canadian producers entered the 1980s in an enviable position: technologically modern, internationally competitive, domestically owned, and profitable. They could win simultaneous plaudits from the nationalist left as a model for economic sovereignty and from Bay Street finance as a sound business investment. Yet by mid-decade their fortunes had turned. This began with a precipitous financial collapse, which pushed the leading integrated steel firms to the verge of bankruptcy. It was compounded by the vigorous expansion of electric steel producers, for whom recent recessions were far less threatening. A new set of corporate players became household names: Nucor and Birmingham Steel in the US and IPSCO, Co-Steel, and Ispat in Canada. After 1994 the industry staged its strongest overall recovery in decades, though this too is rife with contradictions. The gulf between the old and new steel segments continues to grow on both sides of the border, as mini-mills now challenge integrated firms in their most lucrative markets. Furthermore, the big integrated producers remain politically defensive, and the international trade environment is still intensely predatory.

In sum, a variety of political strategies have been forged and applied in the cause of commercially profitable steelmaking. Each can be seen as an exercise in business politics, as steel capitalists vie with other industries and other classes in efforts to harness state powers. Several variables deserve particular attention here: the capacity of domestic steel firms to act collectively in their own interests, the wider business alignments in which the steel sector is embedded, and the forms of politico-business coalitions that mediate the exercise of state authority affecting steel. For the balance of this chapter, we will examine the rich history of steel industry politics, from the nineteenth-century tariff lobbies to the recent financial crises. The roots of such extraordinary transitions can be found in the structure of the steel sector, particularly in the period since World War II. The strategies of two contrasting enterprises,

Stelco and IPSCO, illustrate the fortunes of integrated and mini-mill producers, respectively. This is followed by case studies of two highly revealing political struggles: the Algoma-Sysco struggle over surplus capacity in the steel rail segment and the steel dumping and safeguards battle that has been ongoing since 1997.

Historical Development

As a colonial dominion and a late industrializer, Canada lacked many of the attributes of the nineteenth-century steel powers. Furthermore, it faced a more advanced industrial base to the south. Not surprisingly, these factors played a role in shaping the Canadian development path. It was the opening of the railway age that triggered the initial Canadian investment in steel. Railways were a pre-eminent imperial instrument, offering a superior infrastructure for transporting staple export products such as lumber and grain and for opening the western territories to civil authorities, farming, and mineral exploitation. Over more than half a century, from 1850 to 1911, the Canadian railway boom provided the largest outlet for iron and steel products, as well as improved access to material inputs such as coal and iron ore.

Over time, the embryonic frontier ironworks gave way to larger scale enterprise, combining pig iron smelting and its conversion to steel (Inwood 1987). The National Policy Tariff of 1879 played a role in this, though not in the classic import-substitution fashion. Overall, in fact, the tariff had less impact on steel than on many other Canadian manufacturing sectors, because the early industry was divided on the question of protection and could not present a common front to Finance Minister Tilley in his tariff-setting consultations (Forster 1986). The structure of the industry played a major part in this. There were few firms to speak for tariff protection of basic pig iron products. Moreover, many secondary steel manufacturers sought to import pig iron free of duty, while at the same time gaining tariff levies on their own steel outputs. Tertiary steel users in the finished product industries wanted duty free steel all round. This collective discord opened the way for Tilley to strike his own compromise. In general, the government would consider tariff protection only for products that were being successfully manufactured in Canada. Consequently, most firms tended to specialize and to concentrate their tariff lobbies on particular product lines. This left vast segments of Canadian steel demand to be served by imports well into the twentieth century and still today (Heron 1988).

Another policy tool that found greater favour with steel promoters was the "bounty" and "bonus" program. The former involved cash grants to steel producers on the basis of output volume. The federal government's steel bounties were a critical element in attracting US investment to Canada, particularly after the Laurier government took office in 1896. In addition, Ottawa passed a railway subsidies act in 1900 that made the purchase of Canadian-made steel products a condition of eligibility (Naylor 1975). Provincial and municipal bonuses were popular tools to influence site selection. Free land, local debenture investment, cash investments, special tax rates, and cheap power were all negotiable. Not only did they encourage the installation of new capacity, setting off a provincial and municipal bidding war in the process, but they also facilitated the financing of new works and offered an immediate bolster to profitability. These policies succeeded in substituting Canadian steel for imported product, and the expanding industry included foreign-owned as well as domestic plants. Some of the largest basic steel companies, such as Algoma, Dominion Foundries, and Dominion Iron and Steel, were American in origin (MacDowell 1984). Elsewhere, US equipment and technique was widely licensed to Canadian producers. At this stage, basic steelmaking was highly sensitive to material — coal and iron ore — costs, which meant that plants were located at waterside on the Great Lakes in the case of Stelco and Algoma and in Nova Scotia where Scotia Steel and Disco were established. Canada's first steel ingots were smelted by Scotia Steel at Trenton in 1883 (McCann 1981). Fuelled largely by railway demand, Canadian steel was well-established by the turn of the century.

During the corporate merger wave which preceded World War I, the steel sector was once again transformed (Donald 1915). The largest of the new firms became modern joint stock corporations, with dramatically expanded scales of operations and greater linkages of primary and finishing activities. By 1910 the Steel Company of Canada (Montreal-Toronto), Algoma Iron and Steel (Philadelphia-Toronto), and the Dominion Iron and Steel Company (Boston-Montreal) illustrated the new corporate form. Critical to these massive mergers was the financial backing of Canadian finance capitalists, such as Max Aitken and James Ross, and the larger chartered banks, including the Montreal, Commerce, and Royal. This financial segment both directed the consolidation process and oversaw the results from the boardroom.

Following a very prosperous wartime demand for munitions, the basic steel industry was hit severely by the 1920 recession. This was complicated further by the fact that neither Algoma nor Besco, the newly consolidated Nova Scotia producer, had anticipated the close of the railway- building era and so failed to develop alternative lines. Tom Traves has detailed the elaborate tariff lobby of the 1920s, which pitted this increasingly desperate corporate pair against the more diversified Stelco in pressing Ottawa for tariff support, bonuses, and freight rate adjustment (Traves 1979). Once again deep-seated differences among the leading corporations translated into tariff policy autonomy for the federal government. In 1926 Mackenzie King appointed an Advisory Board on Tariffs and Taxation as a specialized regulatory agency to sort out the competing claims. Significantly, it was only when steel tariffs were politicized by the popularity of the Maritime Rights regional protest movement that Ottawa was spurred to action. However, this was soon overtaken by the onset of the Depression.

The steel collapse of the early 1930s can scarcely be exaggerated. At the nadir in 1932, Canadian producers operated at less than 20 per cent of their rated capacity, and for three months in the spring of 1933 no pig iron was produced in Canada at all. However, conditions began to improve in 1936, and considerable new investment was made over the next three years. At the same time, the level of steel imports declined over a decade from 57 per cent in the late 1920s to 38 per cent in the late 1930s, following the devaluation of the Canadian dollar after the nation left the gold standard in 1931 and the rise of tariffs. The recovery trend was dramatically confirmed during the war economy. As an essential commodity, steel fell after 1940 under the Wartime Industries Control Board, which allocated resources and channelled orders, and the Wartime Prices and Labour Board, which regulated price schedules and wage rates. Significant new capacity was installed in these years, with the assurance of full utilization through the Steel Control Office. With the massive accelerated depreciation offered under the federal tax regime, Stelco opted to finance its own expansion during the war years. In addition to this, Algoma and the Dominion Coal and Steel Company (Dosco) benefited by installing new blast furnaces and open hearths, which they operated for the War Assets Corporation and which they purchased at war's end on concessionary terms.

It was also during the war that a new pattern of trade union politics was confirmed. Even before the turn of century, the more skilled or craft workers formed local unions in Canadian iron and steel plants. The Amalgamated Association of Iron, Steel and Tin Workers grew rapidly in Canada after 1917, reporting 13 lodges and up to 8,000 members at its peak. However, it declined just as quickly during the postwar recession. South of the border, Amalgamated's industry-wide strike of 1919 drew intense resistance from the big US companies, which succeeded in breaking the union (Heron 1988). In 1936, a new approach to pan-industry organization was mounted by the Steel Workers Organizing Committee (SWOC), which was also based in the US. SWOC achieved a breakthrough at US Steel in 1937, and independent unions at the Sydney, Trenton, and Hamilton steelworks quickly chose to affiliate, with Algoma workers following in 1940. By 1944, SWOC had more than 50,000 members in 113 locals across the continent.

However, big steel in Canada mounted fierce resistance, and internal union tensions between socialist and communist tendencies also slowed the organizing effort. Of the big three firms, the industry leader Stelco was the last to recognize the SWOC, which became the United Steelworkers of America (USWA) after 1942. In addition to exploiting political cleavages among the workers based upon ethnicity and skills, Stelco organized a works council in 1935 as a non-union consultative mechanism. In 1942, SWOC took a majority of positions on this council and demanded union recognition. Crucial to the resolution of this issue was a wider steel strike in the same year, which drove the King government to issue the pivotal order-in-council PC 1003. This ushered in the modern Canadian system of union recognition and collective bargaining, and the certification of USWA Local 1005 at Stelco was one of its early results. However, the local's first real test came in the 1946 contract negotiations, which led to a three-month strike. Over time, Local 1005 became a leading part of the USWA District No.6 in Canada, and the Stelco negotiations have normally served as the pace-setter for the industry in the period since (Freeman 1982). At nearby Dofasco, by contrast, the union never managed to sink roots, in large part due to the firm's strategy of shadowing the cross-town union wage rates.

Despite this signal breakthrough in industrial relations, the emerging labour movement and its leadership was still firmly outside the political establishment in both the US and Canada. This is strikingly evident in comparison to the ready legitimacy granted to the corporate elite. Consider the fate of the "Murray Plan" in the US. In 1940, SWOC leader Philip Murray proposed that joint labour-management industry councils be established to plan wartime production. This followed a proposal by Walter Reuther, then a vice-president of the United Auto Workers, to build "500 planes a day" through the comprehensive conversion of auto plants. Of the Reuther Plan, US Treasury Secretary Henry Morganthau declared that "There is only one thing wrong with the proposal. It comes from the wrong source" (Hoerr 1988, 274). A similar fate befell the Murray Plan, although the Roosevelt administration went on to staff its wartime control apparatus with thousands of private-sector corporate managers, as did C.D. Howe in Canada with his "dollar a year" men (Bothwell and Kilbourn 1979).

Though the steel industry expected and feared a postwar repeat of the 1920 recession, the opposite was the case. Canada's postwar boom in steel demand was fuelled by several forces, including new investment by the older steel-using sectors such as railways and farm machinery, the rise of new steel-using industries such as oil and gas and tall building and highway construction, and the deferred demand in consumer commodities ranging from automobiles to appliances (Kilbourn 1961). Despite accurate forecasts by the Steel Controller in 1945, the companies waited until well into the Cold War period before the extraordinary expansion of the 1950s. Stelco's capacity tripled as a result, Algoma diversified into structural steel shapes, and Dofasco became the fourth primary steel producer by installing leading-edge technology with its basic oxygen steelmaking process. This surge in steelmaking capacity was linked also to the opening of new ore bodies in Ontario and Quebec, with Canada becoming the world's fourth largest iron ore producer by 1956. During this long boom, between 1950 and 1969, steel production grew by an average of 8 per cent annually in Canada (Cockerill 1974). Measured by global standards, the Canadian steel industry moved from an inefficient base in the early 1950s to a world competitive base by the early 1980s (Barnett and Schorsch 1983).

The modernizing Canadian steel industry was well matched to postwar trends toward liberalized trade. This was a multifaceted and often contradictory process, involving a number of different policy arenas. At home, the big producers petitioned Ottawa to simplify the steel import tariff schedule (once described as an "archaic clutter") and to raise the tariff rates. During an intensive review by the Tariff Board, the steelmakers faced off against a range of steel-using industries (Canada 1957). The final outcome saw the Diefenbaker government rationalize the number of categories and exempted items but leave nominal tariff levels much the same, testifying to the approximate political equilibrium between steel supplier and user lobbies (Luczak 1985). Ottawa enacted a three-tier rate for British Preferential at 5 per cent, Most Favoured Nation at 10 per cent, and a General tariff of 20 per cent.

A second trade policy track was the tariff negotiations sponsored by the GATT beginning in 1947. Here a different political calculus applied, since steel was only one of many key Canadian industries that stood to be affected by multilateral tariff reductions (Singer 1969). Moreover, national steel markets remained strikingly self-sufficient until the 1970s. By the Kennedy Round GATT Agreement in 1968, the effective level of steel tariff protection among the member states had dropped to about 7 per cent in proportion to the value of the steel produced (Jones 1986). Equally important to the steel sector were the non-tariff rules on such matters as dumping, subsidies, and emergency market protection that evolved in subsequent GATT rounds in Tokyo and Uruguay. In the meantime, Canadian policy balanced the permissible tariff protection, and an anti-dumping defence against unfairly discounted imports, with the advantages to domestic steel users of rapid and affordable access to foreign supplies. Far more than GATT-level tariff rules, the 1965 signing of the Canada-US Autopact provided an historic boost to the integrated Ontario steel producers. As auto assemblers and parts manufacturers repositioned their operations on a continental scale, a new era began for suppliers of sheet and coil steel. Notably, provincial government involvement in regional steel companies also dates from this time (Litvak and Maule 1977). The respective roles of Saskatchewan, Quebec, and Nova Scotia are described in considerable detail in the sections below.

These trade liberalizing tendencies coincided with a major realignment in world steel production patterns (Hawes 1986). Not only had western European countries successfully rebuilt their heavy industrial base during the 1950s, but Japan was about to break into international trade with extraordinary success. Capitalizing on new technologies and scale economies built into new greenfield production complexes, the Japanese achieved exceptional productivity advances (Junkerman 1987). This was particularly evident during the 1966-72 period, when much of their new capacity came onto the global market. As a result, Japan was able to displace the US as the dominant world exporter, while making significant inroads into the US and western European domestic markets.

At the same time, the rising imports triggered a protectionist trade spiral, which in many ways recalled the difficulties of the Depression years. This syndrome began in 1969, when the US State Department concluded voluntary export restraint agreements (VRAs), whereby the European Community and Japan agreed to limit their exports to the US. Since then, the instruments of import control have varied from VRAs to anti-dumping duties to the trigger-price mechanism and compulsory quotas (Stritch 1991). Some of these are discussed in greater detail below. In general, the Canadian industry weathered the turbulent 1970s better than most. Politically, Ottawa became alarmed at the macro-economic impact of steel price increases at a time when stagflation—low growth combined with galloping inflation—had disrupted the postwar economic banquet. In effect, the steel industry was implicated because of its foundational role in so many product chains. Justice Willard Estey's inquiry findings—that neither prices nor profits had risen significantly beyond customary levels—seemed to forestall further regulatory intervention, and Estey's generally laudatory assessment of steel performance echoed a more general perception of steel as a Canadian business success story (Canada 1974). Most producers had taken advantage of federal tax incentives to embark on major modernization programs, which left them very competitive with their US rivals. Indeed, a sense of the steelmakers' relative comfort with the prevailing policy climate can be found in the 1978 report by the tripartite iron and steel committee that formed part of Ottawa's industrial strategy effort. While the task force pointed out business sensitivities to trade, transport, and environmental regulations, there was no sense of crisis or even urgency in its recommendations (Canada 1978c). Ottawa recognized, however, that the major threat to this sector loomed in the international sphere. The centrepiece of federal strategy since 1983 has been to secure guaranteed access to the US market as it has become more aggressively defensive and to shield Canadian producers from the potential backlash (Kymlicka 1987). This forms the subject of the second policy case below.

Only since the 1980s have international pressures caused acute disruption to Canadian producers. Following the 1982 recession, another massive round of capital investment took place, coming on-stream just as the 1990 recession hit. This period was punctuated also by major strikes at Stelco in 1981 and 1990, Algoma in 1990, and the Sydney Steel Company (Sysco) in 1988, which meant new import flows and lost domestic sales. As the only non-unionized primary producer, Dofasco was viewed in business circles as an exemplary model. Its sudden 1989 takeover of Algoma Steel made it briefly the largest Canadian company and fourth largest on the continent. However, this also saddled Dofasco with heavy financial charges, streamlining problems, and the extraordinary decision three years later to write off the Algoma investment altogether. This pushed Algoma into bankruptcy protection and an eventual employee buy-out of the firm in 1992, involving the USWA and the province of Ontario. In Nova Scotia, the province orchestrated one last Sysco "modernization" strategy by phasing out the ancient blast furnaces and scaling down production with a new electric arc furnace. This transformed Sysco from an integrated producer into a mini-mill and paved the way for a series of failed privatization efforts that led ultimately to closure.

For the industry at large, the combination of structural and cyclical dislocations peaked during the 1992 recession. None of the integrated Canadian producers escaped the catharsis. A joke circulating on Bay Street asked "what is the difference between Algoma and Stelco?" The answer was "about three months." Steel sales later recovered along with the business cycle, drawing most companies back into profit by 1994, and five buoyant years followed. When the next profit crunch hit in 2000, it was largely a pricing slump driven by record steel imports and desperate price discounting in response. Clearly, the industry continues to face a number of intractable challenges, as is evident below.

Structure

In understanding the performance of firms in the market, analysts often look to the structure of the industry for clues. The modern study of industrial organization works with a continuum of market structures, ranging from the highly decentralized form in which a plethora of firms compete under conditions that approximate Adam Smith's notion of perfect competition, to the highly concentrated form in which a few large firms or a single monopolist dominate a product market. Each form suggests a different relationship between firms and different dynamics in the economic processes of price-setting.

In the case of steel, the early nineteenth-century stage of charcoal iron production, when factories were small and markets were local, may have been the most *dispersed* form of enterprise. However, the greatest degree of *competition* likely occurred toward the end of the century, when demand for iron was mounting and entrepreneurial capitalism was at its height. Once primary steelmaking began in Canada in 1883, the scale and complexity of production moved beyond the grasp of most family entrepreneurs. Scotia Steel was forced to sell shares across the province and beyond to fund its blast furnace-steel furnace-finishing mill complex (Frank 1977). By 1920, three more or less integrated industrial enterprises dominated the manufacture of basic or primary steel—Stelco, Algoma, and Besco. It was only after World War II that the Hamilton-based Dofasco eclipsed the Sydney and Montreal-based Dosco.

Figure 3-1 illustrates the various production processes involved in integrated steelmaking. The basic materials include coal, iron ore, and limestone. In the classical method, the coal is baked in coke ovens, after which it fuels the blast furnaces in which iron ore is transformed into pig iron. This molten iron is then turned into steel by one of two processes. Prior to 1950 it was generally cooked in open hearth furnaces, but after that time the basic oxygen furnace proved markedly superior. In 1954 Dofasco was the first North American producer and second in the world to adopt the basic oxygen furnace method. While integrated North American firms were relatively slow in switching to this method, it had become the industry standard by the 1980s and 1990s. Traditionally, the molten steel was then poured into ingots to solidify for storage or transportation. These would later be reheated and processed further in forging presses or rolling mills. As the name implies, the superheated ingots would be rolled into basic shapes such as billets (long and thin), blooms (wide and thick), and slabs (wide and thin). Each of these shapes provide intermediate materials for additional milling. Billets are fashioned into bars, rods, and

Figure 3-1

A Flow Line for Steelmaking

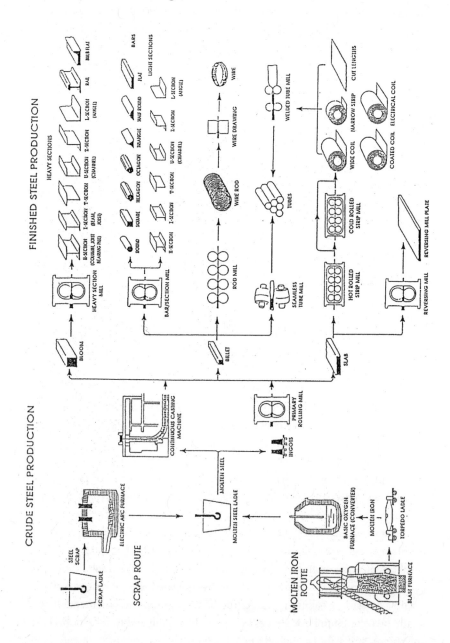

narrow pipe, while blooms are turned into structural shapes, rails, and wider seamless pipe. Slabs are fashioned into plate, welded pipe, steel sheet, and strip, with the last two rolled in either hot or cold forms.

The second technological breakthrough following the war involved the continuous casting of molten steel. By eliminating the considerable handling, storage, and energy costs associated with ingots, this offered major efficiency gains to integrated producers since the basic shapes could move directly to processing mills. Once again it was a Canadian firm that paved the way, as Atlas Steel became the first North American continuous caster. Of course each company or establishment fashions a particular version of these sequences and in this way acquires the particularities that set it apart. For example, Dosco operated blast furnaces and open hearths in Sydney, Nova Scotia, while many of its rolling mills were in Montreal, where steel ingots were shipped.

The chief alternative to integrated steelmaking involves the melting of steel scrap or directly reduced iron in electric arc furnaces. As they were initially far smaller in capacity than open hearth and basic oxygen furnace units, the electric arc plants became known as mini-mills. Dependent only on a ready supply of scrap metal and cheap electricity, these mills could be located in dispersed regional markets rather than clustered around the mine and shipping networks of the Great Lakes basin. Until recently, mini-mill product lines were also simpler, concentrating on small rods and reinforcing bars and small diameter pipe and structural shapes. However, recent breakthroughs in mini-mill technology, particularly in the field of thin slab casting, have erased many of the previous boundaries. The most dramatic instance of this occurred in 1989, when mini-mill leader Nucor Inc. opened its first thin-rolled strip mill at Crawfordsville, Indiana. This put Nucor in direct competition with the high-end products of the integrated sector. It has since been joined by a score of similar plants with annual capacities exceeding one million tons.

Except in the United Kingdom, where the nationalization of the 14 largest firms in 1967 created the near-monopoly, British Steel Corporation, most modern steel sectors have been loosely oligopolistic. In such cases, pricing behaviour and business strategy assume a distinctive form. The high costs of both capital and technology create barriers that discourage new entrants. At the same time, there is little incentive among the established giants to engage in competitive pricing, since all firms tend to move in a parallel direction. A price war simply serves to curtail profit margins without significantly altering market shares. The more common practice is for a leading firm, such as US Steel or Stelco, to supply price leadership, while other companies shadow it with parallel schedules (Adams and Mueller 1982).

The political implications of oligopoly structures in steel have been significant. In the US following World War II, the executive branch was continually
sensitive to the exercise of market power by the steel giants. On the one hand
was concern that the oligopoly could restrain output in order to push up
prices. On the other was an equal concern that steel was becoming an engine
of inflation, as record-setting wage settlements were reflected in the rising
prices charged to domestic steel consumers. As president, Harry Truman
went so far as to threaten to build new state-owned steel plants to create
added capacity and diminish the prospects for cartel behaviour. A decade
later, John Kennedy staged a showdown with the president of US Steel over a
posted price increase at a time when Washington had appealed for restraint to
curb inflation (McConnell 1963). More generally, the US government adopted
a liberal approach to primary steel performance in the postwar period. As
Borrus puts it, Washington granted autonomy to "US corporations to pursue
their own strategies for adjusting to changing conditions of competition in the
international marketplace" (Borrus 1983, 61).

In Canada, the relationship between steel capital and the state has been
far more accommodating. An explanation of this begins with corporate
strategies. Unlike those in the US, Canadian firms have never concentrated
on supplying a complete range of product groups. Rather, they have specialized in commodities that offer significant advantages of scale in serving the
domestic market, while allowing imports to fill the demand for less profitable
products. Dofasco, for example, concentrates on flat-rolled products and
plate, while Algoma has focussed on structurals, rails, and plate. IPSCO has
specialized in pipe and tubular goods, and Atlas is a stainless steel producer.
There has also been geographic specialization, reinforced by the significant
role of freight rates in marking up delivered prices. Stelco comes closest to the
US integrated model of wide product mix (Litvak and Maule 1985). Between
1950-80 Canadian firms were extremely successful in this approach. Plant
expansions were geared to serving average rather than peak cyclical demand,
and therefore capacity utilization was high.

Structurally, Canadian steel has been a relatively tight oligopoly—the big
three integrated firms accounted for over 70 per cent of raw steel output in
1980 and the CR_4 in 1982, based on sales, was 78 per cent. Yet even this does
not capture the degree of market segmentation by geographical and product
area. Among primary producers in the domestic market, inter-firm rivalry has
not been strong. Neither, until recently, did steel imports inject the expected
level of competitive pressure, since they too tended to serve segments that
were poorly contested by Canadian producers. As for exports, which flow

almost entirely to the US, levels were modest up to about 1980, when they doubled as domestic firms grappled with the twin realities of new capacity and declining domestic demand.

On balance, the government of Canada has been strongly supportive of the steel industry strategy. Indeed the years from 1950–80 were marked by a very durable "policy-strategy consensus" between Ottawa and the producers (Litvak and Maule 1977). This complementary package of policy measures consisted of several elements. In acknowledging the efficiencies of the corporate allocation of product markets, Ottawa opted to mute its application of the anti-combines law and overlook the potentially anti-competitive nature of segmentation. Similarly the long-term steel supply mix of domestic and imported products, which relied upon imports to fill product gaps rather than to pose head-to-head competition, was facilitated by the structure of the customs tariff schedule. Finally, Ottawa's policy of accelerated tax write-offs for capital investments served to encourage plant modernization, process innovation, and timely asset replacement (Masi 1991).

For both industry and government, the postwar strategy eroded during the 1980s, and the policy environment changed markedly. The most revealing indicator was the rising size of the traded steel market, both imports and exports. This went beyond the usual oscillations of rising Canadian imports in strong business cycles and Canadian exports to customers south of the Great Lakes. In Canadian steel, the capacity expansion in the late 1970s, coupled with the energy-driven recession of the early 1980s, left firms in desperate need of new markets, and for this they looked to the US (Ontario 1988). The sharp rise of Canadian exports to the US, which doubled between 1984 and 1986, soon drew the attention of US producers and their political allies in Washington. In effect, Canadian firms had exploited their exemption from Washington's 1985 import restraint program to fill the space vacated by the controlled offshore suppliers. Anticipating a more aggressive US response, Ottawa moved in 1987 to establish a Canadian steel monitoring program. Aimed at accurate tracking of Canadian exports, this obliged any firm shipping more than $5,000 worth of product to the US to secure a permit. The era of contested and managed steel trade had arrived (Rhodes 1993). In 1992, Canadian firms were named in a new flurry of US anti-dumping actions, and a cross-border steel trade war broke out. These events underline the critical significance of the comprehensive Canada-US trade talks for domestic steel. In effect, Ottawa's new steel-sector strategy was based on bilateral trade policy.

This, of course, cut both ways. After 1985, steel imports into Canada jumped from their traditional 20 per cent level to as high as 40 per cent. In this new pattern, cheap offshore supplies began to penetrate Canada in the full set of product segments, including those most important for domestic producers. This forced the latter into systematic defensive actions before the Canada Customs and Revenue Agency and the CITT. At stake was not just lost sales but sharp downward pressures on the price front. This differed qualitatively from the traditional practice of allowing imports to service non-core segments or cyclical peak demand. Moreover, with steel-using industries and steel traders mobilizing to defend this potential advantage, the steel politics of the 1990s centred increasingly on the CITT.

Over time, the rise of the mini-mills served to transform the shape of the industry. In Canada, electric arc steelmaking accounted for one-third of total output by the 1990s. In the US, the total was closer to one-half. Even with the multi-plant tendency of leading mini-mill operators like Nucor and IPSCO, the inherently innovative and dynamic nature of this sector injected important levels of competition. Moreover, as the mini-mills moved out of their original fields of strength (wire, rebar, and small pipe) toward higher value segments like thin-rolled slabs, plate, and coil, their competitive significance was further magnified.

There are also more than 50 steel *processing* companies in Canada, most of which operate on a smaller scale and under more competitive conditions. Theirs is a different world from the "big three." Most are involved in secondary steel manufacture or "finishing" operations, and they are as likely to buy as to make their material inputs. This is the production of higher value-added outputs, often for narrow but lucrative niche markets. Note that the secondary steel manufacturers seek the lowest cost inputs and for that reason may welcome primary steel imports, even under conditions which suggest dumping.

In a more general sense, Magaziner and Reich stress the importance of distinguishing between an *industry* and the multiple *businesses* that may exist within it. Here a business is defined by its markets, technologies, production methods, cost structures, and barriers to entry. Within the steel industry, for example, the businesses include basic commodity steels as above; specialty steels, such as stainless and tool steel; and specific finishing operations, such as high tolerance steel. The key point, essential both to business strategy and public policy, is that "the basis for competitive productivity improvements varies significantly across various businesses" (Magaziner and Reich 1982, 155) within an industry. Consequently, the industry, or even subindustry, is not always an adequate unit of analysis.

More pertinent can be the major product line, which is the basis of manufacture, pricing, and supply. For purposes of this analysis, six leading pricing

Figure 3-2

US Steel Basis Prices

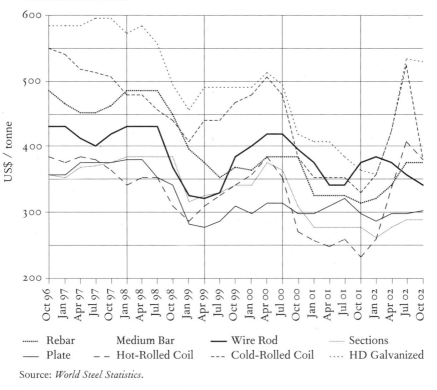

Source: *World Steel Statistics.*

categories are distinguished. Blooms and billets can be fashioned into heavy and light sections for basic construction and building materials. Billets are also the basis for the merchant bars and reinforcing bars used in concrete structures and drawn wire rod and tube/pipe products. Steel slabs can be rolled into plate products as well as hot-rolled and cold-rolled strip. The recent patterns of pricing in the US market are captured in Figure 3-2. The cyclical shifts since 1996 are clearly evident.

Strategies

> *Steel Company of Canada (Stelco Inc.):*
> *Profile of an Industry Leader*

Stelco was founded in 1910 as the culmination of a massive merger of Ontario and Montreal steel and fabricating companies. It brought together the basic steelmaking capacity of Hamilton Iron and Steel, along with a number of Ontario wire, screw, and nut and bolt manufacturers, and the pipe, wire, and nail facilities of Montreal Rolling Mills. The transaction was overseen by the finance capitalists serving on both boards. As a modern joint stock company run by professional managers, it was a prototype of the integrated steel corporation (Kilbourn 1961). Until recently, it ranked among the top ten producers in North America.

After World War I, Stelco emerged as the industry leader in a highly concentrated Canadian market. Its traditional position accounted for about 40 per cent of raw steel capacity, and it acted as the undisputed price leader among the integrated producers. It was at Stelco in 1946 that the USWA made a signal breakthrough in Canada, as Local 1005 was recognized under Ontario labour law (Freeman 1982). During the 1950s the central Canadian producers reached a price- and market-sharing understanding which persisted for some time. Stelco's predatory pricing strategy, strenuously applied in the face of new competitors such as IPSCO on the prairies, triggered an anti-combines investigation in the 1960s.

Throughout this period, Stelco was strongly committed to plant modernization, underwritten by the accelerated federal capital cost allowances for investment in steelmaking equipment. However, in its approach to new technologies, Stelco's record was more timid and somewhat blinkered. While its Hamilton competitor Dofasco introduced basic oxygen furnace technology to Canada in the 1950s, Stelco remained committed to the older open hearth furnace method until the late 1960s. In the other major technological innovation, the continuous casting of semi-finished steel shapes, the mini-mill Atlas Steel ran ahead of the larger integrated mills including Stelco. Nonetheless, during the long postwar boom period, when domestic supply was expanding and the threat from imports was low (totalling less than 5 per cent of the domestic market annually), Stelco was a blue-chip fixture for any corporate investment portfolio.

By the early 1980s, the postwar golden age was gone forever (Corman, *et al.* 1993). Stelco entered the decade with several newly modernized facilities among its 15 Canadian plants. Among these was the very important greenfield facility of the Lake Erie Steel Company at Nanticoke, Ontario, which reflected state-of-the-art technology (Sandberg and Bradbury 1988). However, such improvements came at the cost of an extremely high debt load. The situation was accentuated by a lengthy 125-day strike in the fall of 1981, which not only cost the company sales and market share, but also opened the way for heightened import penetration. As an economic recession began to build in the wake of the labour settlement, Stelco was forced into a program of labour cuts, which would average 1,000 layoffs per year over the next decade. In 1982, the company announced the largest loss in its history. No dividends were paid to shareholders for the next three years, and earnings fell below 1 per cent in each of the subsequent three years to 1987. This coincided with a renewed US import restraint campaign, which (as described below) touched Canada for the first time. Plans to build the company's entire future around the Nanticoke plant were significantly modified by the decision to modernize the Hilton works, a less expensive option by one-half billion dollars.

In January 1988, Stelco announced a corporate reorganization that saw its business regrouped into two new divisions. Stelco Steel handled the primary steelmaking operations while Stelco Holdings coordinated the finished manufacturing operations. The former included not only the integrated works at Hilton and Lake Erie but also electric arc operations that produced bar and billet products at Contracoeur, Quebec, and Strathcona, Alberta. The latter included the Z Line Company for zinc-coated galvanized steel. Developed as a joint venture with Mitsubishi Canada Limited and the Baycoat partnership with Dofasco, Z Line turned out pre-painted flat-rolled steel. Stelco Holdings also included pipe and tubular facilities at Welland, Ontario, and Camrose, Alberta, as well as wire, rod, and nail plants in Hamilton.

No one in the industry was prepared for the rapidity or the depth of the financial decline suffered by the Ontario integrated producers, including Stelco. Stelco's large refit was completed in 1990. However, even with these improvements, it was calculated that the Hilton works faced a $30 per ton cost disadvantage in semi-finished steel. Stelco took this burden into the business cycle downturn that began in 1990. The company's production was shut down during the three-month strike that autumn, while industry imports to Canada rose from 16 to 25 per cent. Stelco's financial loss for 1990 set a new record at $196m. Table 3-3 details the impact of this slumping demand on the company's net income over the next four years. Soon Stelco was vigorously shedding assets, such as a share of its Camrose pipe plant ($40m) and a Kentucky coal mine ($30m). Even so, a new share issue was successfully placed in the summer of 1991. However, Stelco's crippling debt load of more than $1b added to the difficulties of cyclical downturn and import penetration. By 1992 the situation was catastrophic. Stelco suspended dividend payments to shareholders in March 1992, while the Canadian Bond Rating Service declared Stelco's debt, once a solid triple-A, to be subinvestment grade. The company's shares, which had traded at $27 as recently as 1987, sold for an incredible $0.93 in the early part of 1993.

Table 3-2

Stelco and IPSCO Distribution of Output, 2000

Stelco Inc.		IPSCO Inc.
30%	Cold-rolled and coated	0
28%	Hot-rolled	44% coil and plate 21% cut to length
5%	Plate	
16%	Rod and bar	0
8%	Wire products	0
8%	Pipe and tubular	21% energy 8% non-energy 6% large pipe
5%	Other	0

Source: Calculations by author based on corporate reports.

However, with the economic recovery boosting the auto-sector demand for steel, Stelco moved back into profit in 1994, and its shares had recovered to more than $9 by that September. The next five years were a period of mounting demand and improved balance sheets, which allowed Stelco to retire debt and augment operating cash while also undertaking significant capital investment for gains in operational efficiency. In 1997, new Stelco President James Alfano underlined the buoyancy, declaring that "these [i.e., today] are the good old days in terms of North American steel demand" (Alfano 1997). Like Dofasco and other integrated producers, Stelco moved to diversify its product base and to tilt volumes toward the high-value end. In practical terms, this meant emphasis in galvanized and coated sheet products for the auto trade, where big mills still held commanding positions, while also maintaining a strong position in oil and gas tube and piping and developing new-generation construction materials. Table 3-2 illustrates Stelco's product mix at the turn of century.

Trade policy has been a central concern in Stelco's commercial strategy for the past few decades. Traditionally, the company exported about 15 per cent of its output to the US, a flow that has been crucial to high capacity use and profitable operations. Former President Fred Telmer minced no words on the inadequacies of the FTA and NAFTA when it came to fair trade. Their failures to agree on common subsidy rules, he argued, led to the permanent politicization of the US import trade (Telmer 1996a). This was doubly vexing since, despite the refrain of US complaints about offshore "subsidies," the US remains one of the leading subsidizers, with government support in the form of tax breaks, direct grants, and discounted electricity supplies. An additional complication, this time on the Canadian side of the steel market equation, accompanied the rising levels of imports in the 1990s. Stelco joined other firms in voicing frustration with the slow and rigid procedures governing unfair imports to Canada. The practice of "dump and run" importing in which one-shot shipments of cheap steel were sold was well documented, and the damage was inflicted before the trade authorities in Ottawa could respond. Once such imports reached sufficient scale, as in the period after 1997 (discussed in detail below), they forced domestic producers into price discounting as well. Stelco and its rivals argued persistently for more vigorous anti-dumping remedies to match those applied abroad.

Stelco's high profit years between 1994 and 2000 coincided with a booming business cycle. The company initiated a major investment program for the Hamilton and Lake Erie works, spending over $850m and making the latter mill the most efficient plant on the continent. In many ways, the company's central operational challenge was to coordinate the relationships between these two integrated facilities, the brownfield Hilton site in Hamilton, and the greenfield Nanticoke site at Lake Erie, for maximum value. After 1998, Lake Erie's capacity increased by more than one-third, and by 2003 it had surpassed Hamilton in integrated semi-finished steel production. These capital improvements were directly related to labour cost reductions as well, with Stelco's total workforce falling by 24 per cent after 1996.

However the year 2000 saw a sharp negative turn in steel markets, with most major product group prices falling by as much as $100 per tonne. Stelco's thin profit of 2000 was followed in 2001 by the deepest loss for more than a decade. In 2002 a modest net income of $14m suggested a tentative firming; however, any gains were promptly cancelled in early 2003. The first quarter results showed a loss of $44m or $0.45 per share. The second quarter report was significantly worse, with a loss of $82m or $0.83 per share. Certainly, many of the familiar problems had a hand in this: weakening auto sales and output and the continuing open market for low-cost steel imports, with Ottawa refusing to follow the US lead in imposing special tariff measures. To this should be added the dramatic rise in value for the Canadian dollar (more than 20 per cent against the US dollar), which undermined both spot sales and long-term contracts with US customers.

Unfortunately, this was no consolation for Stelco shareholders, and the failure to effectively deal with investor interests led to a dramatic change of management regime in the summer of 2003. For months, Stelco shares had been under pressure as the financial situation worsened. Common shares that traded in the $4 range late in 2002 dropped into the $3 range in the spring of

2003 and to a mere $1 by June. This drop of 75 per cent in less than a year should have signalled the need to reassure beleaguered investors. The traditional conference call with major investors and analysts, which follows the release of quarterly financial results, offered one such opportunity in July. By all published accounts, however, Stelco President Alfano's performance in that call was "disastrous" (Gould 2003). His responses to key questions were seen as hostile and dismissive, and, even more critically, no strategic adjustments were announced. The markets were quick to indicate their loss of confidence, and an investor crisis clearly loomed.

Two days later, Alfano announced his resignation, and Stelco chairman Fred Telmer was named interim CEO. Two days after that, Stelco's prairie mini-mill, AltaSteel, was put up for sale, signalling a willingness to consider restructuring as a response to the 2003 predicament. Telmer took pains to stress the company's secure credit base, with $173m in combined cash and credit available for the transitional period. However, before the end of his first month in office he also announced a comprehensive review of Stelco's business operations. Core assets—chiefly the two integrated plants—would be streamlined and strengthened, while non-core assets would be divested. In November 2003 a new president was appointed. Courtney Pratt was a choice from outside both the company and the industry. Though he had served on the Stelco board of directors for almost two years, his senior management experience was with Noranda Inc., a mining corporation, and Toronto Hydro. The year closed with a far more urgent commitment to realign Canada's largest steelmaking firm. Time, however, was not on Stelco's side. It was forced to seek bankruptcy protection in January 2004.

Interprovincial Steel Corporation (IPSCO Inc.):
From Regional Mini-mill to Multinational

In 1956, the Saskatchewan provincial government approached a group of private promoters to build a steel pipe mill in Regina. The social democratic Co-operative Commonwealth Federation party was in power, and the premier was the charismatic Tommy Douglas. The steel project was part of the government's economic diversification program, which sought to promote forward economic linkages based on primary resources. The western oil boom had begun a decade earlier with the Leduc strike in Alberta in 1947, and the petroleum industry was a massive consumer of steel casings and pipe, both for exploration and transport. With the province's financial support, and the prospect of major sales to the new gas supply network run by the state-owned

Saskatchewan Power Corporation, the Prairie Pipe Manufacturing Co. Ltd. was formed. Three years later an electric arc steel mill was installed, again with private bond capital guaranteed by the state. While the eastern financial centres showed much scepticism at this experiment in mixed public-private enterprise, the business plan proved sound. The steel- and pipemaking operations were later merged and the first profits of the renamed Interprovincial Steel Corporation (IPSCO Inc.) were announced in 1962.

Though IPSCO was then a small regional upstart, it could be seen as a symbol of business and political challenge. It represented indigenous prairie enterprise in contrast to the big firms of central Canada, a heavy industry rather than a light consumer goods venture, and a product of social democratic as opposed to business-oriented politics. IPSCO was one of a number of small steel producers to carve out profitable niche markets, defined both by region and by product line. The specialization strategies of the big integrated firms, discussed earlier, left considerable room for mini-mill operators to capitalize on lower delivery costs to offset higher unit of production costs (Litvak and Maule 1977). While this might suggest the possibility of local and regional monopolies, the big three producers vigorously defended their regional market shares. Stelco had earlier bought up Premier Steel Mills of Edmonton, the first electric arc mini-mill established on the prairies in 1955. A fierce price war in pipe and tubular products followed between IPSCO and Stelco. Throughout this period, the government of Saskatchewan maintained a 28 per cent equity share in IPSCO, along with options on sufficient treasury shares to prevent the company's takeover by an Ontario-based rival.

Prior to 1968, IPSCO concentrated on small-diameter pipe products of less than 16 inches. Then the installation of spiral welding technology opened the way for larger diameter manufacturing of up to an 80-inch diameter pipe. The next decade saw dramatic expansion of output and revenues, together with a range of corporate acquisitions and a broadened equity base. Both Slater Steel of Hamilton and the government of Alberta took 20 per cent stakes in IPSCO, matching that of the government of Saskatchewan. The balance of IPSCO equity was publicly traded. Several acquisitions in Alberta and British Columbia extended the firm's network of pipe manufacturing facilities. All of these factors reflected IPSCO's emerging status as a "prairie" regional enterprise. In the mid-1970s, company sales in western Canada amounted to 80 per cent of the total, with 12 per cent in central Canada and 8 per cent in the US. While most of IPSCO's raw steel production was absorbed by its pipe mills, the company resold primary steel products such as hot-rolled sheet and plate, which it imported from Japan.

A major concern of any mini-mill involves a reliable and affordable supply of scrap metal. Regionally, IPSCO developed a supply contract with the Saskatchewan Wheat Pool, whose local elevator agents (1,300 of them in the 1970s) bought and collected steel scrap. In addition to prairie supplies, IPSCO bought scrap from the north-central US states ranging from Montana to Wisconsin and also operated an auto shredding unit in Regina. Still, recycled metal supplies tended to be volatile, and a scrap surcharge formed an element in mini-mill pricing strategies. In addition, the cross-border scrap market was subject to government export controls. All of these factors encouraged firms such as IPSCO to explore alternative metal sources. One possibility was sponge iron. Already utilized by Sidbec-Dosco and Stelco's Quebec works, it was created by direct reduction from iron ore. This meant that significant impurities remained, although the iron content was approximately 90 per cent. IPSCO was actively investigating direct reduction during the 1970s (Litvak and Maule 1977). More recently, direct reduction iron-making has played a major role in the growth of mini-mill operations in the US Gulf states, with the metal being imported mainly from the Caribbean.

IPSCO's relations with its provincial government shareholder have tended to be positive though not problem-free. An agreement to develop an iron mine in central Saskatchewan in concert with Denison Mines and the province was terminated in 1971 when the newly elected New Democratic Party government rejected the deal. However, that same government recognized IPSCO's strategic economic role by signing a 1974 economic agreement with Ottawa to facilitate iron and steel diversification. The program included support for iron ore exploration, preparatory work toward a direct reduction plant, diversification of primary and secondary mill operations, and infrastructural investment.

The company's annual turnover (or sales), which had doubled over the 1960s, increased ten-fold during the 1970s to more than $250m. During the 1980s this doubled again to half a billion dollars. However, all was not clear sailing. By 1981, IPSCO was producing and rolling more than half-a-million tons of steel per year for manufacture at six pipe mills. Yet as a pipe producer, IPSCO's fortune was tightly tied to the notoriously volatile petroleum sector. Consequently, the world oil price slump, coupled with the broader business recession of 1982, cut deeply into company performance. Just prior to that time, Roger Phillips, formerly of Alcan Aluminum, had been recruited as CEO. In spite of the company's enviably low debt-to-equity ratio of .38:1, the slump brought layoffs, plant closures, and a program of severe cost-cutting the following year (Verburg 1998).

Table 3-3

Stelco and IPSCO Net Income, 1988–2001

Stelco Inc. C$	Year	IPSCO Inc.	
		C$	US$
+ 99.7 m	1988	+ 28.8 m	
+ 93.8 m	1989	+ 19.3 m	
− 196 m	1990	− 4.7 m	
− 136 m	1991	+ 36.3 m	
− 127 m	1992	+ 15.5 m	
− 36 m	1993	+ 28.7 m	
+ 115 m	1994	+ 57.7 m	
+ 156 m	1995	+ 81.7 m	
+ 79 m	1996	+ 83.0 m	
+ 137 m	1997	+ 132.0 m	
+ 119 m	1998	+ 113.0 m	
+ 107 m	1999		+ 74.3 m
+ 4 m	2000		+ 57.7 m
− 178 m	2001		+ 38.9 m
+ 14 m	2002		+ 20.3 m

Source: Stelco and IPSCO Annual Reports.

Once the business cycle rebounded, another major modernization was launched. By mid-decade, IPSCO was installing continuous casting technology and a new heavy plate line. This joined a product mix that already included flat-rolled steel (of up to one-quarter inch), large diameter pipe and tubes, and reinforcing construction steel. In an era when the Canadian steel industry was entering increasingly treacherous waters, IPSCO's record of shrewd strategy, efficiency, and profitability made it a darling of the financial markets. A comparison of the financial results of Stelco and IPSCO (Table 3-3) illustrates why this was so.

For some years, however, the uncertainties of Canada-US trade policy had posed concerns to IPSCO management. Along with the rest of the Canadian industry, it had supported Ottawa's initiative in the 1985-88 free trade negotiations and in the subsequent battle for ratification. However, IPSCO President Phillips was one of the first Canadian steelmen to question whether secure market access had in fact been achieved by the terms of the deal. Moreover, when adverse trade rulings (discussed in detail below) began to surface in the late 1980s, it was time for a strategic reorientation. Under the circumstances, the surest way to ensure access to the US market was to circumvent the dumping and countervail danger by producing south of the 49th parallel. The year 1988 proved pivotal. IPSCO purchased pipe manufacturing works in Nebraska and Texas and a distribution and processing centre in Minnesota. It also bought Western Canada Steel from the Cominco mining firm of British Columbia. Overall, capital acquisitions cost more than $36m that year. The US presence was extended in 1990 by the purchase of an Iowa pipe fabricator. By 1991 IPSCO operated plants in four mid-western US states as well as the three western provinces. Its 1992 results showed a capacity utilization of 78 per cent and a modest profit, in striking contrast to the massive losses being reported by the Ontario industry in the depths of the new recession.

However, the boldest stroke was yet to come. As we have seen earlier, Nucor tackled the integrated steel sector head-on by opening its hot-rolled sheet plant in Indiana in 1989. The crucial breakthrough was in thin slab casting. This dramatically reduced the rolling costs and allowed Nucor to compete in markets which had been, until then, big steel's last redoubt. In 1993 IPSCO signalled plans to emulate Nucor by building a major greenfield facility in the US mid-west. This triggered a bidding war among several US state governments, which proposed financial subsidies, attractive plant sites, and right-to-work laws designed to discourage trade unions. Ultimately IPSCO chose to build in Montpelier, Iowa, which offered prime Mississippi River access for material movement and close proximity to the Illinois-Ohio heartland. When it became fully operational in 1998, this plant had a twin-shell electric arc furnace with a rated output of 1.25m tons per year, a continuous casting machine, and a flat-roll mill for plate (Metal Bulletin 2000). This served to more than double IPSCO's raw steelmaking capacity, while diversifying its product line into growth markets such as plate and coiled steel (see Table 3-2). It also allowed forward-links to the tube and pipe mills at Geneva, Nebraska, and Camanche, Iowa. For 1997, IPSCO's operating profit per ton of steel produced—a key performance indicator—was reported to be the highest in North America, averaging $139 (*World Steel Outlook* 1998).

IPSCO's US investment grew by another quantum leap in 2001 with the opening of Montpelier's sister mini-mill at Mobile, Alabama. With the addition of the Mobile plant, IPSCO's total crude steel capacity rose to 3.5m tons annually. As Table 3-1 reveals, IPSCO thereby overtook the traditional number three firm, Algoma, in the league tables of Canadian steel capacity.

In only a decade IPSCO had accomplished a breathtaking strategic shift while remaining profitable throughout. By 2001 company shipments originated

Map 3-1

IPSCO Inc. Major Plants, 2002 — by Location, Product, and Scale

Note: IPSCO also operates five cut-to-length and temper mills in Canada and the US.

65 per cent in the US and 35 per cent in Canada (IPSCO 2001). Moreover, while 35 per cent of sales were pipe and tubular products, a full 65 per cent were in steel coil and discrete plate. The range of sourcing and marketing possibilities had been dramatically extended. While oil and gas products remained an area of competitive strength, the company was no longer overly dependent on a single sector. IPSCO sold flat products as commodities as well as processing them. Since rolling capacity exceeded that for raw steelmaking, the company also purchased low-cost hot-rolled sheet from other US and foreign suppliers, for highly profitable onward processing in IPSCO plants. A case in point is the 300,000 ton per year tube and pipe mill constructed at Blytheville, Arkansas, adjacent to Nucor's mini-mill that supplies the steel input.

Despite this confirmed binational profile (captured in Map 3-1), IPSCO's head office remains in Regina, and its Canadian operations remain an important source of corporate earnings. It seems clear, however, that its future investment and growth opportunities are overwhelmingly in the US. In short order, IPSCO has become a leading member of the US mini-mill sector. With such massive recent capital outlays, however, comes a challenge. As the new plants take up the commercial weight, IPSCO's business must be carefully managed if the company is to maintain its place in the top tier of profitable North American mini-mills. The problems experienced in commissioning the Montpelier works, that resulted in delayed testings and start-up, and the legal disputes with the contractor, Mannesman, offer a case in point.

One area where IPSCO has become increasingly sensitive is the import question. Where the company was once insulated by its product mix and the natural defences of a prairie regional fiefdom from cut-priced imports, it is now directly threatened by the flood of flat product imports. Operating on both sides of the border has strongly shaped IPSCO's perspective on the import trade question. In order to protect this business, the company has resorted to trade remedies against discounted imports, as conditions required. Beginning in the early 1990s, IPSCO joined Algoma, Stelco, and Dofasco in the anti-dumping actions against US imports during the Canada-US steel war. Since 1997, it has been active again in rolled sheet and coil actions against a range of off-shore producers. In the US, IPSCO has joined the mini-mill sector in calling for a multilateral agreement on capacity reduction and sub-sidy rules. As such, Phillips has argued that emergency safeguard actions "at home" should be viewed not as a solution in themselves but "as tools to bring the rest of the steel producing world to the table to negotiate a lasting solution to global overcapacity and oversupply" (Phillips 2001).

Policy Issues

The Sysco-Algoma Rail Dispute: Inter-firm Conflict in a Situation of Surplus Capacity

During the mid-1980s, a fascinating public dispute broke out between the Ontario integrated producer, Algoma Steel, and one of its rivals, the Sydney Steel Company (Sysco) of Cape Breton, Nova Scotia. On the surface, the dispute was about rival bids to supply steel rails to the transportation sector. The two protagonists claimed to be fighting for survival in a segment of the steel sector which, so the argument went, had room for only one of them. However, a closer examination reveals that this conflict sprang from deeper issues of corporate repositioning, foreign trade prospects, and the role of state enterprise in regional economic policy. It also reveals the complex interplay among nominally distinct public policy questions in the "steel sector." Just as the rail dispute sprang from a deeper well of controversies, it triggered, in turn, a number of collateral developments. This interplay of issues is illustrated in Figure 3-3.

The immediate precipitant was the February 1986 signing of the Canada-Nova Scotia Sub-Agreement on Steel Modernization, according to which the two governments would share the costs of a major restructuring at the Sysco works. From the outset, this was resisted by rival rail producer Algoma, which was completing a rail mill upgrade of its own. Six months later the premier of Nova Scotia, John Buchanan, sought to further enhance Sysco's prospects by demanding a guaranteed rail purchase agreement with the federally owned Canadian National Railways (CNR). This resulted in a highly visible provincial lobby to Ottawa and an equally intense Algoma counter-lobby. For a time, it also threw the future of the modernization agreement into question. While the province was successful seven months later in obtaining a form of rail purchase guarantee, this came at a cost. The question of state "subsidies" to Sysco gained high political visibility, engendering policy consequences of its own. Before long, this triggered a cross-border trade dispute over Canadian rail shipments to the south. After rulings by the US Department of Commerce and the ITC, Sysco rails were subject to a 100 per cent countervail, which effectively shut them out of the US market. Within months of the preliminary ruling, the premier began ruminating about a Sysco privatization. Not all of these issues will be fully canvassed in detail here; however, the phenomenon of issue-interdependency should be recognized as a potential feature of any micropolitical setting.

The roots of this conflict run far deeper than the Sysco refit, however. It centres on a product with a powerful historical past in Canada, encompassing all phases of the product cycle. It was in the 1870s that steel rails began to supplant iron as the industry standard (the last iron rails in America were produced in 1884). Where iron rails were prone to breakage and required replacement as frequently as every six months, steel rails could last as long as a decade. As we have seen, production for the railway sector was the backbone of the entire steel industry prior to World War I. Yet with the massive railroad corporate mergers that followed 1917 (Perl 1994), this demand slumped severely, and it was destined to remain modest between and after the two world wars. The railway sector accounted for 20 per cent of Canadian steel demand in 1949, 10 per cent in 1959, and only 6 per cent in 1966. Certainly metallurgical advances continued, with harder rails permitting heavier loads and more challenging route designs, but whether they were destined for the domestic or

Figure 3-3

Policy Interplay on Steel Rail Dispute

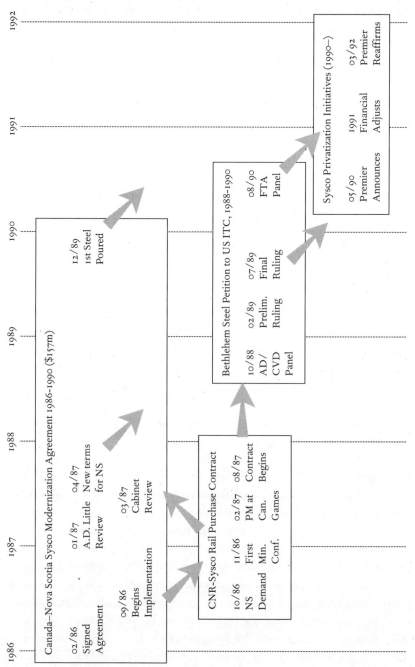

export markets, the demand for steel rails was largely for replacement material rather than for new lines.

Yet despite this relative decline, neither Sysco nor Algoma was willing to abandon its rail mill. In fact, each company had excellent reasons for persisting. The likelihood of new entrants was slight; each firm possessed production and marketing expertise; and with the right product mix, there were prospects for export sales to supplement domestic demand. Equally important, both firms underwent changes of ownership and strategy during the postwar boom period, which deepened their commitments to rail products. Brief accounts of the two rail competitors will be helpful in understanding the controversy that follows.

ALGOMA STEEL AND CANADIAN PACIFIC

With its main mills located at Sault Ste. Marie in northern Ontario, Algoma Steel was historically the second largest primary producer in Canada. It owned iron mines in Ontario, coal mines in West Virginia, and a limestone supplier in Michigan. Many of these inputs were delivered by its Central Ontario Railway, which consisted of rail, marine, and trucking groups. On the death of Sir James Dunn in 1956, control of Algoma passed to the Mannesmann steel and manufacturing conglomerate of Germany. Algoma's main product lines included semi-finished shapes—blooms, billets, and slabs—as well as rails, structurals, and plate. In 1973, control of Algoma returned to Canada, as the Canadian Pacific Industries conglomerate added the steel works to its diversified manufacturing portfolio (Goldenberg 1984). This opened another intra-corporate connection, since the Sault steel mill sold most of its rail output to the holding company's crown jewel, Canadian Pacific Railway (CPR), one of the two dominant national carriers.

By the mid-1980s, rails were the smallest of Algoma's product lines, amounting to a mere 7 per cent of output value in 1984 and 5 per cent in 1987. Shifts in rail demand were threatening even this modest business. The rail market extends from the traditional carbon steel rails, now considered a low-end product, to intermediate grades, and finally to premium grades of head-hardened rails. Rails made of carbon steel had certain limitations, particularly in high-stress locations on grades and turns, in severe climates, and on sections where longer and heavier loads were the trend. Increasingly, Canadian railways demanded the superior, head-hardened product. Significantly, there were no Canadian producers of head-hardened rails, a gap that both Algoma and Sysco had plans to fill. In the meantime, both CPR and CNR met their growing needs for hardened rail by imports, principally from Japan, while

Table 3-4

Rail Purchases by Company, Year, and Product

Buyer	Year	Carbon Rails	Hardened Rails
Canadian Pacific	1982	65,000 t[a]	510 t
	1986	46,000 t	25,000 t
Canadian National	1982	164,000 t[b]	18,000 t
	1986	44,000 t	36,000 t

[a] Algoma supplied 65,000 t.
[b] Sysco supplied 88,000 t.

the domestic rail manufacturers scrapped over the shrinking market for the other grades. The rapid penetration of imports and displacement of domestic carbon product can be seen in Table 3-4.

It was this growing imbalance that Algoma sought to remedy in the capital modernization strategy that followed the 1982 recession. In 1986 it opened a new facility for the manufacture of long 78-foot (or 24-metre) rails, double the previous length, in both carbon and intermediate grades.

The timing is crucial here, since it coincides closely with the finalization of the federal-provincial modernization program for Sysco. Algoma was sufficiently concerned to protest, in its 1985 Annual Report, the "potential government subsidization of a domestic competitor's uneconomical products, in an already fully supplied market" (Algoma Steel Inc. 1986, 3). For Algoma the next step was to install head-hardened rail machinery. While engineering work was announced in 1986, this investment was deferred the following year in a move that was publicly linked to the Sysco modernization, which also aimed to install head-hardened capacity. Though Algoma recorded a profit in 1987, it was the first in six years, and the company's overall financial position remained precarious. Total sales were recovering from the 1982-83 recession, but remained short of the peak year in 1981. Before continuing the rail mill upgrade, Algoma trained its sights on blocking the Sysco modernization.

DOSCO-SYSCO AND THE CNR

Owned by a series of British syndicates, Dosco was the least dynamic of the big-four firms when it came to postwar expansion and modernization. If Hamilton's Dofasco emerged as a technology leader in the 1950s, then Dosco was clearly a laggard. In most product lines, it was a residual supplier of rolled steel, doing best when domestic supply was tight. Its predicament worsened in the 1960s as world supply began to overtake world demand. Dosco's mill workforce, which peaked at 4,000 employees in 1961, declined by 25 per cent over the next five years. Traditional Dosco segments such as light shapes and wire rods became increasingly import-sensitive. However, one area of continuing strength was in light and heavy steel rails.

The situation was compounded by a major historical anomaly. Dosco's coking, pig iron, and steel ingot facilities, together with bloom and billet, rod and bar, and rail mills, were located at tidewater in Sydney, Nova Scotia. The site was originally dictated by its proximity to local coal and to Newfoundland ore supplies. However, Dosco's main rolling and finishing mills were located 1,000 miles farther west in Montreal and Contracoeur, Quebec, far closer to the central Canadian market. So long as this bi-polar configuration persisted, Dosco could not even consider the emerging technologies such as continuous casting, which offered substantial economies by eliminating several intermediate stages on the road to semi-finished steel.

On 13 October 1967, Dosco suddenly announced the closure of its entire coal and steel complex at Sydney. Threatening the very core of industrial Cape Breton, this "Black Friday" decision shook the province. After a period of protracted negotiations, the governments of Canada and Nova Scotia stepped in to purchase the coal and steel works, respectively, to ensure that heavy industry would continue. The coal mines were vested in the federally owned Cape Breton Development Corporation (Devco) while the steel mills went to the provincial Sydney Steel Corporation (Sysco). Commercial interlocks between a network of state corporations were critical to their mutual viability. For Devco, this meant coal supply contracts to Sysco and, in later years, to the Nova Scotia Power Corporation. In addition, Sysco remained a primary supplier of steel rails for CNR, the federal state enterprise. At various times this procurement relation was mandated by the federal government, as in the five-year accord struck in 1977.

The Sydney works had been in provincial state hands for almost two decades before the rail controversy surfaced. Achieving a renaissance at Sysco was always going to be difficult, given the facilities, the product mix, and the times (Bishop 1990). With the Montreal rolling mills absorbed in 1967 by the Quebec state company Sidbec, the truncated character of Sysco—rooted in primary steelmaking and semi-finished shapes—was accentuated. Moreover, after the initial acquisition, the government of Nova Scotia took the position that any further Sysco modernization must be funded internally from earnings (Nova Scotia 1968). Yet by running the plant flat out in the early years—record production was achieved in the early 1970s—managers risked breakdowns and quality slippage. In fact, the acumen of Sysco executives was constantly being questioned. One attempt to install continuous casting and basic oxygen furnaces was bungled badly (Surrette 1975). In 1974, the Nova Scotia auditor general was extremely critical of Sysco managers and directors for poor capital cost reporting and control provisions (Nova Scotia 1975). Certainly, the calibre of Sysco management was open to question, as no fewer than seven senior executive teams had held office by 1989.

For a brief period in the mid-1970s, Sysco's structural frailties led the Nova Scotia government to consider the alternative of a world-scale greenfield plant. Sited south of Sydney at Gabarus Bay, it would be owned by a consortium of Canadian and Japanese or European firms. The plan was to concentrate on intermediate steel products for world markets at competitive prices, certainly a debatable strategy in times of mounting global surpluses (APEC 1974).

While the full story has never been traced, one formidable political obstacle to this plan lay in rival provincial ambitions. The government of Quebec and its crown steelmaker Sidbec also sought a greenfield plant to better balance the province's production and consumption of steel, since at that time up to 30 per cent of Quebec's 1.6m tons of rolled steel consumption was filled by imports. Yet another greenfield proposal was under active discussion on the west coast, linking the government of British Columbia and Nippon Steel (Litvak and Maule 1977). Interprovincial commercial rivalries are not uncommon in Canadian business or politics. Yet in the wider federal context of the 1970s, particularly with the rise of the Quebec sovereignty movement, a political stand-off ensued at the cabinet level in Ottawa. Then Stelco announced plans for its new greenfield project at Nanticoke, Ontario, on the shores of Lake Erie. By 1977 the window had closed on the Gabarus project, though not before postponing the Sysco modernization issue for a further five years. Almost as a consolation prize, Ottawa agreed to join Nova Scotia in funding a $23m incremental Sysco upgrade during 1976-79, called Sysco I, and a subsequent modernization program of $96m from 1982-85, called Sysco II. One blast furnace was rebuilt, while the bloom and rail mills were renovated.

Throughout this period the steel mill continued to post annual losses ranging from $20m to $50m. In an effort to break this cycle, Sysco management advanced a radical but highly contentious proposal in the mid-1980s. It would install an electric arc furnace and a universal mill as the basis for a more efficient but smaller scale steelmaking operation. This could supply the rail mill (and any diversified lines) with far cheaper raw steel. In effect, Sysco would make the unusual transition from integrated steelmaker to mini-mill. The

Sysco workforce of 1,250 could be cut by as much as half. In spite of resistance from USWA, this strategy formed the basis of a third intergovernmental deal, the Canada-Nova Scotia Sysco Modernization Program Agreement signed early in 1986. This four-year project called for the investment of $157m, of which two-thirds came from Ottawa. Almost $87m was allotted for a 150-ton electric furnace, while $28m covered new rolling mills, $15m went for caster and transfer projects, $10m for a ladle refining system, and $15m to renovate the bloom reheat furnace. Prior to this program, Sysco had the capability to manufacture carbon and intermediate steel rails. With the completion of the modernization, it would be able to produce head-hardened rails as well.

THE RAIL SUPPLY CONTROVERSY

The CNR rail contract controversy erupted in September 1986, as the Sysco III Agreement was due to begin. In one sense it was a replay of the long-standing practice of forging policy links between state enterprises in search of both commercial and partisan advantage. Credit for such deals has long been claimed by regional politicians. In 1977 CNR was directed into a five-year contract with Sysco for rail supplies, a move widely attributed to Allan MacEachen, the Nova Scotia regional minister in the Trudeau Liberal government and a skilled political patron. The question resurfaced during the dying weeks of the Nova Scotia provincial election campaign of November 1984. Elmer MacKay, the Mulroney Conservative government's regional minister for Nova Scotia and Minister of National Revenue, promised a new CNR purchase contract with Sysco.

This, however, proved difficult to deliver. Not only was the business policy orientation of the Ottawa Conservatives at odds with the principles of state enterprise and politically administered contracts, but MacKay faced a regional rival at the cabinet table. The solicitor-general, James Kelleher, represented the Algoma district of Sault Ste. Marie, where he had earlier served as mayor. Thus, each mill and region had a legitimate claim for regional economic development support. Faced with mutually offsetting pressures, the federal government focussed instead on modernization assistance to Sysco since a precedent had been clearly established by the 1976 and 1982 agreements. In Ottawa the lead agency was the Department of Regional Industrial Expansion, whose negotiations with Nova Scotia throughout 1985 culminated in a December draft announcement and the February 1986 signing described above.

Once the agreement had received Treasury Board approval, an engineering contract was awarded in September 1986 to SNC of Montreal. In triumphal spirit, Premier Buchanan marked the event by declaring that "no one can now say the future of Sysco isn't secure." Yet only a month later the premier raised this very prospect by suggesting that, since the February signing, declining steel markets were jeopardizing the agreement and might even lead to plant closure. Crucial in this respect was an updated Sysco market study, required under the agreement and conducted by the Arthur D. Little Company. Evidently Buchanan feared that a negative appraisal might open the way for cancellation. By mid-November, the premier announced that Sysco required a ten-year guaranteed contract to supply 85-90 per cent of CNR's rail needs. Furthermore he declared that he would press Prime Minister Mulroney on the issue at an up-coming first ministers conference. In the event, Buchanan was blunt. He told the prime minister that the commercial viability of CNR needed to be subordinated to the needs of maritime regional development (Arnold 1986). Other ministers joined this debate. In one revealing outburst, one Conservative Nova Scotia minister declared that if Allan J. MacEachen (a Liberal) were still in cabinet, "we'd have five years [worth of CN rail orders]" (Jeffers 1986). The obvious premise underlying this sudden campaign was that, lacking additional supports, the Sysco business plan was insufficient for the changing steel market. As the year closed, Ottawa declared that no decisions could be taken before the Little Report was considered. In effect, the contents of the marketing study held the potential to reopen or even cancel the Sysco modernization agreement.

For several reasons, the rail purchase contract was not the simple panacea that the Nova Scotians claimed. Major policy initiatives originating beyond the steel sector played a key role. The Mulroney government was committed to deregulating the transport sector, with the prospect of extensive rail-line abandonment to follow. This major initiative was being driven by the minister of transport, John Crosbie, who encouraged the railway to buy the cheapest rails available, regardless of source. Neither was CNR willing to be burdened by expensive procurement guarantees. President Maurice LeClair stated that CNR would buy rails according to need but would not stockpile surplus inventory (MacDonald 1986a). Equally, the Sysco case was beginning to draw attention in the sensitive circles of Canada-US cross-border trade talks, where its status as a heavily subsidized state enterprise was quickly flagged. Coming in the autumn of 1986, when the Reisman-Murphy free trade talks were well advanced, this raised additional concerns about jeopardizing Canadian access to the US steel market. In short, the rail purchase guarantee, and even the Sysco agreement itself, were opposed by a new and widening group of political players in Ottawa.

Clearly the policy orientations in Ottawa and Halifax could not have been more different. This served to limit the impact of federal ministers, such as MacKay, as Sysco advocates (Hoare 1986). After lambasting CPR for ignoring Sysco products in the past, MacKay attempted to shrug off his 1984 election promise as "said out of conviction" but "a bit too draconian," while suggesting that an exclusive supply relation with CNR was not essential for Sysco (MacDonald 1986b).

A matter of public discussion by January 1987, the much-anticipated Little Report was sobering though not fatally damaging to Sysco's prospects. Forecasting that deregulation would permit CNR and CPR to abandon half their routes, it scaled back the forecasted Sysco rail demand by 25 per cent, from 127,000 tons to 88,000 tons per year (Arnold 1987). This included an assumption that Sysco would obtain 90 per cent of CNR's business, a key provision for the company's viability. Little also suggested that more aggressive export sales might compensate for this drop in domestic sales.

While the Nova Scotia government refused to publicly release it, the market study circulated widely within government and industry, and it triggered an intensified lobby by Algoma. The company complained about the state underwriting of surplus rail capacity at Sysco, at a time when Algoma had already invested in a similar plant. Furthermore, it argued the inevitability of US countervail rulings, once the Americans found the heavily subsidized Sysco to be guilty of unfair trading practices. Finally, it predicted that Algoma's planned upgrade into head-hardened rail-making would be jeopardized if the Sysco plan went ahead (CP News 1987). Though quickly dismissed in Nova Scotia as a self-serving central-Canadian petition, Algoma's case raised some cogent points. It claimed the mantle of private-sector efficiency, established the inevitability of surplus rail capacity, questioned Sysco's realistic access to the US market, and highlighted the threat of additional Sysco capacity in other product markets.

If nothing else, this contributed to the difficulty of resolving the multi-layered deadlock. Indeed, it was not until mid-February that the terms of a CNR rail purchase policy were announced. This took place immediately prior to the opening of the Canada Games in Sydney, where the prime minister faced the prospect of widespread picketing and protest. Effective in August 1987, Sysco would be assured of its "traditional share" of CNR purchases, subject to cost and quality standards. This was set at 80 per cent for carbon and intermediate rail products and for head-hardened rails when available. However, the deal did not specify a minimum tonnage, and it rejected the option of CNR stockpiling rail inventory. With the expected hyperbole, Buchanan declared the future of Sysco to be secure for 35 to 40 years (Campbell 1987).

No sooner had this news settled than it appeared the modernization agreement was again under threat. In mid-March, Kelleher announced that a five-member cabinet committee had been struck in Ottawa to review the agreement. Largely unnoticed during the rail purchase battle was the fact that a revised Sysco business plan needed formal political approval before the modernization could begin. For Department of Regional and Industrial Expansion Minister Michel Côté, this review was initially described as routine. Yet according to Kelleher, the Little Report insisted that Sysco required more diversified production than rails alone (MacDonald 1987). Critical to this was the $125m universal mill that would enable Sysco to manufacture structural shapes, another staple product at Algoma.

In confirming this view, Côté also opened a new spat with Halifax. He stated that Nova Scotia must meet certain obligations of its own, by writing off Sysco's accumulated debt and guaranteeing the purchase of the universal mill, before the modernization agreement could be approved (MacDonald and Arnold 1987). Part of his aim may have been to force the province into supplementary commitments. It may also have been prompted by credibility problems in the business plan. For example, the press reported that Sysco proposed to finance the universal mill internally, based on cash flow, a most improbable prospect for such a marginal operation. Buchanan resisted linking either the debt write-down or the universal mill to the modernization agreement, arguing that each was a separate issue to be dealt with in phase three of the Sysco revival.

THE BETHLEHEM TRADE COMPLAINT

During October 1988, Bethlehem Steel lodged a complaint in Washington, alleging unfair trading practices by *both* Canadian rail producers. As the long-time number two firm in the US primary steel industry, Bethlehem was experiencing all the problems of big steel. This included the habit of lashing out with unfair trade complaints against a panoply of foreign competitors, particularly when recession brought lags in domestic demand. For some it seemed puzzling that a giant such as Bethlehem would trouble itself with the small share—amounting to only $10m in 1987—which Canadian producers held of this relatively minor product line. It can be better understood as a strategic move in the wider field of steel duties and quotas. With the US's five-year steel quota deal up for renewal in 1990, Bethlehem was offering an opening gambit and warning that it intended to see Canadian steel imports included in the new import regime.

The Bethlehem complaint was a two-part action, alleging that the Canadians were dumping rails—in effect, selling in the US at prices below the costs of production—and that they were also benefiting from state subsidies, such as capital grants, cheap energy, and wage subsidies, which permitted lower manufacturing costs and price structures. To maximize the impact of its argument, Bethlehem lumped the Canadian rail producers together, illustrating dumping with the case of Algoma and subsidies with Sysco. Bethlehem also sought to demonstrate a mounting threat from Canadian rails, which were said to have risen from 1 per cent of the US market in 1985 to 12 per cent in 1987.

In February 1989 the Commerce Department issued its preliminary rulings, dismissing the subsidy case against Algoma but assigning a 103 per cent provisional duty against Sysco's 5,000 tons, which represented 0.5 per cent US market share. Since the effective price of Sysco rails had been doubled, it ceased to ship into the US market. On the dumping charges, a very nominal duty of 2.7 per cent was assigned against both firms, suggesting that the level of the "dump" was almost insignificant (Romain 1989). Algoma stressed that this was the second consecutive unfair trading charge it had parried in the US system.

Five months later the final rulings were announced. While Sysco's countervail duty was reduced to 94 per cent, the anti-dumping duty was revised significantly upward to 39 per cent. On the matter of material injury, the ITC split on a vote of three to three. Since a majority is required to overturn the earlier decision, the effect was to confirm the final duties (CP News 1989). Both Canadian firms announced their intention to appeal the Commerce decisions under the Binational Dispute Settlement provisions of the FTA, which had only recently been ratified. A year later the panel, which could only review the decision against US statutory law, voted four to one to uphold the countervail against Sysco and also upheld the dumping finding.

Occurring at a time when Sysco was undergoing its final modernization to install the new electric furnace and universal mill, this was a major setback. Since the subsidy attributes were now structurally embedded in Sysco's operations, there seemed no way to reopen US access, where Sysco had hoped to achieve a 15,000-20,000-ton share of the market by, it claimed, displacing European and Japanese importers. For Algoma the dumping findings were equally troubling, since it had traditionally held a larger share of rail exports to the south.

THE ROAD TO PRIVATIZATION

The first steel was poured from Sysco's electric furnace in December 1989. One month later the government of Nova Scotia moved to assume Sysco's

$785m long-term debt at the same time as its rail order book began to expand. Perhaps in the hope of capitalizing on this positive business image, Premier Buchanan announced, quite unexpectedly in May 1990, that Sysco was a candidate for privatization (Jeffers 1990). In the event, little private interest was shown. Throughout this period, the company continued to record losses ranging from $20m to $50m per year, and despite the technical success of the new facilities, its finances were terminally ill. Its provincial line of credit was exhausted early in 1992, and it was forced to turn to the chartered banks for a $30m credit line secured by the mill machinery.

It was after the change in provincial leadership that privatization was again launched, this time with conviction. Perhaps the government that had fought so hard for modernization and rail contracts could never fully give up on Sysco. However, with John Buchanan appointed to the Senate of Canada, and Donald Cameron succeeding him as premier, the way was clear for a new look at the steel plant. Early in 1992 Cameron announced that Sysco was losing $4m per month, a cost that the province would no longer bear (Henderson 1992). An investment agent was named to facilitate privatization. Years passed with one prospective deal collapsing after another. Ultimately the electric furnace was closed, and Sysco was wound down without sale in 2000.

RESOLUTION

The rail dispute grew out of two rather particular complaints raised by Algoma as objections to federal public-sector "subsidies" to Sysco. It will be evident, however, that these were tangible hooks on which to hang more general complaints about Ottawa's response to mounting surplus capacity in primary steel. The first point involved Sysco's use of federal modernization grants to enter new product lines in which Algoma had an ongoing stake. The second concerned CNR's purchasing practices in securing steel rails, from which Algoma claimed it was being arbitrarily excluded by the politically imposed tie between the railway and Sysco. The third was the danger of trade-disrupting retaliation by US firms and authorities sensitive to subsidized Sysco products. Significantly, all these issues were raised toward the end of a major wave of capital investment, when all the major companies had fallen into loss. In an atmosphere of mounting sensitivity to "unfair trade" issues and when the FTA had just been concluded, Canada was drawn into US import quotas for the first time.

Once the Gabarus greenfield option was gone, the government of Nova Scotia groped uncertainly for a compromise between full closure and wholesale renewal. Through the terms of the Sysco subagreement to the Canada-Nova Scotia Economic and Regional Development Agreement, the two governments provided capital for upgrading selected facilities in the plants, and Halifax injected additional support of its own for other projects. This was negotiated in the context of a down-scaling strategy, which closed the blast furnaces and open hearths in favour of an electric arc mini-mill. Not only did this involve the reduction of the Sydney workforce to less than 1,000, but it also begged the question of how successfully a brownfield conversion of this magnitude could be accomplished under public-sector management and in a glutted market. As an additional wrinkle, the spectre of eventual privatization hung over the conversion. Thus, the state agencies would cover the cost of downgrading and would assume the company's accumulated debt in order for private capital to purchase a newly modernized facility.

In these disputes, the pattern of politicized steel rivalry shifted from Nova Scotia-Quebec to Nova Scotia-Ontario. The ambitious greenfield hopes of the 1970s dissolved into the reluctant capacity phase-outs of the 1980s. The stronger firms achieved this through market capitalization, while the weaker were forced into close collaboration with state authorities. Despite considerable ambivalence, Ottawa concurred in the support of Sysco and in financing its restructuring. Significantly, this was the mirror image of the modernization strategy being pursued by the three Ontario firms, this time with the support of federal tax expenditures rather than direct subventions. However, the nature of the changes in an increasingly globalized steel economy served to constantly threaten both the Sysco project and the wider prospect for an industrial policy for steel. Indeed, it is striking how limited the policy tools of the 1960s and 1970s proved to be in dealing with the industrial policy problems of more recent times. A further instance of this syndrome can be seen in the case that follows.

Steel Trade Politics: Dumping and Safeguards, 1997-2002

The past decade has witnessed two distinct eras of steel trade politics in Canada. Through most of this, the industry has succeeded in maintaining a consensus position, although the policy discourse and strategic focus of that consensus has shifted significantly. The first phase took place against the backdrop of the continental free trade negotiations—the Canada-US deal of 1989 and the NAFTA of 1993. Canadian steel was in the forefront of domestic manufacturers supporting these pacts for the sake of guaranteed access to the US market. From this perspective, the failure to achieve either a binding dispute settlement mechanism or a code defining fair and unfair subsidies left the outcome unfinished and flawed. Stelco President Fred Telmer made this point repeatedly in the mid-1990s, pointing to the Australia-New Zealand Free Trade Area as a better model of what was needed (Telmer 1996b). By mid-decade, however, it was clear that Washington would never agree to such measures and that, flawed or not, the deal was done.

This dawning awareness, coupled with the mixed Canadian experience in defending its US market access before the Department of Commerce and ITC, led Canadian steelmakers to revise their policy agenda. It now centred on defending the Canadian domestic market against unfair imports. In effect, this amounted to a call for Canada to adopt a US-style position of aggressive reciprocity through a strengthening of Ottawa's import policy regime. The first step was the 1996 review of the Special Import Measures Act (SIMA), which Finance Minister Paul Martin justified as giving Canadian producers equivalent protection against unfair imports to that already offered to Americans. The Canadian Division of the USWA described this challenge succinctly as "making our end of the playing field as bumpy as theirs" (USWA 1996).

According to the CSPA, the revised SIMA was "instrumental in correcting some of the worst market disruptions arising from the [1998] Asian crisis and [1999] Russian economic problems" (CSPA 1999). At the same time, it continued to press for improved *implementation*, through new SIMA regulations and new CITT rules. The complaints included the slow pace of securing injury determinations and remedies, the opportunities still available for "dump and run" damage, the failure to deal effectively with importers engaging in "source switching" to evade injury rulings, and the general reluctance of the CITT to aggressively pursue its mandate of policing fair trade. On these matters, the CSPA advanced a number of solutions. It wanted Ottawa to initiate fast-track investigations when new importers sell at levels already found injurious in earlier cases, where import volumes surge quickly, and where supplying nations have a history of subsidizing. It also sought the levy of duties retroactive to the point of first import, rather than only the "go-forward" basis following an injury ruling. Finally, a system of enhanced import monitoring was proposed to place greater reporting burdens on importers to demonstrate that they follow fair pricing policies in cases of major shipments (CSPA 2002). To these points, the USWA added the call to streamline CITT procedures on findings of injury with legislated definitions of "material injury" and "domestic market." Not only would this speed results and avoid certain masking manoeuvres by traders, but it would also curb the discretion of CITT officials in applying what the USWA considered to be neo-liberal policy values.

This shift of orientation coincided with a sharp rise in the year-over-year levels of steel imports to Canada after 1997. Among the contributing factors were the Asian financial slump, the increasingly desperate efforts of post-communist states to earn hard foreign currency from dumped exports, and the surprisingly robust North American business cycle. Asian steelmakers, reluctant to curtail production in the face of declining domestic demand, were aggressive discounters in Western markets. Eastern European steelmakers were far more interested in earning foreign exchange than profit. Although Canadian steelmakers invested more than $5b in the 1997-2002 period, steel-using industries' demand for inputs outpaced the capacity of domestic producers to fill. As a result of these compound factors, steel imports to Canada peaked at more than 40 per cent of apparent consumption in the year 2000. Swamped in excess supply, North American prices plummeted sharply. Dofasco President John Mayberry described the 2001 business year as the "lowest ebb" in a decade (Keenan 2002). In the integrated sector, this proved to be too much for many marginal players. Among the high-profile US bankruptcy protection candidates were three of the big six firms: Bethlehem, LTV, and Wheeling-Pittsburgh. In Canada, Algoma followed suit.

Throughout this period, steelmakers defended against imports with the familiar tools of anti-dumping and countervail petitions. Over the five years beginning in 1997, Canadian firms launched 12 anti-dumping cases against a broad spectrum of foreign suppliers. Ten of these proved successful and provided levels of relief for up to five years. A summary of these cases, highlighting the central interests and outcomes, appears in Table 3-5 below.

However, these contingent protection measures were far from perfect in neutralizing unfair trade. Despite the apparent strong trend in winning injury remedies, Canadian producers viewed the CITT's performance as mixed. Since only the strongest cases tended to be raised in the first place, the industry has a high expectation of securing meaningful remedies in every action. It was particularly vexed by the two "no injury" rulings of 2001, reached in spite of Canada Customs and Revenue Agency findings of dumping margins of 24 per cent on cold-rolled steel and 15 per cent on galvanized steel. This prompted the steel producers to appeal the CITT rulings to the Federal Court of Canada (Alfano 2002). Furthermore, international merchants and import agents have proven adept at "source switching" or juggling suppliers of dumped products to replace those named in past cases. As a result, the chronic problems of oversupply and weak pricing continued irrespective of particular anti-dumping victories or defeats. Moreover, the political leverage exerted by steel-trading merchants and steel-using industries was substantial. Not only

Table 3-5

Steel Dumping/Subsidy Cases Before the Canadian International Trade Tribunal, 1997-2001

Case No. and Product	Imports From	Domestic Suppliers	Product Market	Finding / Date
NQ 97-001 Hot-rolled carbon steel plate	Mexico; China; South Africa; Russia	Algoma; Stelco; IPSCO	$ 500 m	Threat of injury 27/10/97
NQ 98-001 Stainless steel round bar	Germany; France; India; Italy; Japan; Spain; Sweden; Taiwan; UK	Atlas Specialty Steels	$ 30 m	Material injury 04/09/98
NQ 98-003 Stainless steel round bar	Korea	Atlas Specialty Steels	$ 30 m	Material injury 18/06/99
NQ 98-004 Flat hot-rolled carbon and alloy steel sheet	France; Romania; Russia; Slovak Republic	Stelco; Dofasco; IPSCO; Algoma; Ispat-Sidbec	$ 2.8 b	Material injury 02/07/99
NQ 99-001 Cold-rolled steel sheet	Argentina; Belgium; Russia; Slovak Republic; Spain; Turkey; New Zealand	Dofasco; Stelco, Ispat-Sidbec; Algoma	$ 1.1 b	No injury; threat of injury 27/08/99
NQ 99-002 Concrete reinforcing bar	Cuba; Korea; Turkey	Co-Steel; Ispat; Stelco; AltaSteel; Stelco-McMaster; Gerdau-Courtice; Gerdau-MRM; Slater	$ 290 m	Material injury 12/01/00
NQ 99-004 Carbon steel plate	Brazil; Finland; Ukraine; India; Indonesia; Thailand	Algoma; Stelco; IPSCO	n/a	Material injury 27/06/00
NQ 2000-002 Stainless steel round bar	Brazil; India	Atlas Specialty Steels (Slater)	n/a	Material injury 27/10/00
NQ 2000-007 Concrete reinforcing bar	Japan; Latvia; Moldova; Poland; Taiwan; Ukraine	Stelco; AltaSteel; Co-Steel; Gerdau-Courtice; Gerdau-MRM; Ispat; Slater	$ 350 m	Material injury 01/06/01
NQ 2000-008 Corrosion-resistant steel sheet	China; India; Malaysia; Russia; So.Africa; Taiwan; India (subsidy)	Dofasco; Sorevco; Stelco; Continous Colour Coat Ltd.	$930 m	No injury; no threat of injury 03/07/01

Table 3-5 (continued)

Steel Dumping/Subsidy Cases Before the Canadian International
Trade Tribunal, 1997-2001

Case No. and Product	Imports From	Domestic Suppliers	Product Market	Finding / Date
NQ 2001-001 Flat hot-rolled steel sheet and strip	Brazil; Bulgaria; China; Taiwan; India; Korea; Macedonia; So.Africa; Ukraine; Yugoslavia; Saudi Arabia; NewZea.	Stelco; Dofasco; Algoma; IPSCO; Ispat-Sidbec	$ 3.3 b	Material injury (excl. Korea, So.Africa, Malaysia) 17/08/01
NQ 2001-002 Cold-rolled steel sheet	Brazil; China; Taiwan; Macedonia; Italy; Luxembourg; Malaysia; Korea; South Africa	Dofasco; Stelco; Ispat-Sidbec	$ 830 m	No injury 09/10/01

Source: Canada. 1997/98 to 2001/02. Canadian International Trade Tribunal, *Annual Reports*.

did they participate in the dumping and subsidy cases, but their interests were acknowledged in the conceptual and technical underpinnings of the import trade bureaucracies. This is explored in more detail below.

After the turn of the century, Canadian steel producers sought a closer and more clientelistic relationship with federal ministers and agencies. This is not entirely new, as there has been administrative collaboration on steel trade monitoring since the mid-1980s. However, as the decade of the 1990s progressed, the emphasis shifted from export access to import protection. It assumed urgent proportions with the US steel trade crackdown of 2001.

THE BUSH IMPORT CRACKDOWN

June 2001 marked a turning point in the alignment of steel-sector interests on both sides of the border, when US President George W. Bush initiated an ITC investigation into whether rising imports were causing serious injury to the US steel industry. In trade policy terms, this was known as a "safeguard" investigation, permitted by the GATT (now WTO) general agreement in cases where sudden surges of imports threaten to impose long-term injury on domestic producers. In the US case, such actions are authorized by section 201 of the US Trade Act of 1974.

Several features distinguished this steel safeguard review. First, since it was ordered by the president, it signified policy concern at the highest executive level of the US government. Second, it was prompted by a wider set of political forces than producer complaint alone. As Hufbauer and Goodrich (2001) have pointed out, the president needed to reassert control of a steel policy agenda that was increasingly being claimed by Congress. Though none have yet been enacted, congressional bills to impose quantitative rollbacks on steel imports to mid-1990s levels have been passed by the House of Representatives, along with loan guarantees to troubled firms and excise taxes on steel sales to fund legacy costs. In the face of presidential inaction, these measures stood to gain increasing support. Moreover, Bush needed to show strength in defending domestic industrial interests if he hoped to win congressional consent for fast-track trade approvals for both the Doha round of the WTO negotiations and for the Free Trade Agreement of the Americas. Certainly, this was not the first time that Americans used aggressive trade remedy measures to lever concessions in international negotiations (Hart 2002).

It is important to recognize that safeguard investigations, in both the US and in Canada, differ qualitatively from the anti-dumping and countervail duty petition process. First, the safeguard process can be global, examining the full spectrum of steel products and markets, where the petition process is incremental and normally concentrated on particular products, firms, and time frames. A safeguard package can accomplish at a stroke what might be impossible to achieve in a decade of bottom-up unfair trade complaints. Second, the package of safeguard remedies that is ultimately adopted can seek a political balance across the range of products, points of origin, and domestic producer-merchant-consumer interests. Given the existence of prior trade treaties such as NAFTA and the WTO agreement, any safeguard remedies must be consistent with these rules as well. Third, the conceptual and decision-making criteria that drive safeguard inquiries are distinct from those of anti-dumping and countervail duty actions and make possible different solutions and outcomes (detailed below).

Without question, the Bush safeguard initiatives of 2001-02 altered the political climate for steel trade globally. In the short run, it forced all industrial interests with stakes in the US market to participate in the ITC hearings. It also signalled that import restrictions of some sort were imminent, since the president could hardly unleash such extensive public expectations without delivering a result. However, the actual outcome remained open to contest by steelmakers, steelworkers, steel users, US state agencies, and foreign governments. Even though they inevitably carried technical and even moral force, the ITC conclusions were, in the end, only recommendations to the White House. It was up to President Bush to decide the binding terms of the safeguard package.

Here the president had the opportunity to fashion a broader steel-sector policy that linked import restrictions to wider solutions. He could choose, for example, to craft a financial strategy for the massive legacy costs that blocked the industrial reorganization of big steel. Or he could fashion a wage insurance program to cushion the dislocation of workers as integrated companies downsized or amalgamated. Or he could insist that meaningful capacity reductions be achieved by eliminating non-competitive steel enterprise as a precondition for any form of sectoral relief. Finally, he could build a US steel strategy that levered the other major global steel regions—the European Union, the post-communist states, and the Asian bloc—into comparable capacity cuts through a broader WTO or OECD (Organization for Economic Cooperation and Development) trade agreement package.

THE STEEL SAFEGUARD QUESTION IN CANADA

For several years, Canadian steel firms had been pressing Ottawa for support beyond the standard anti-dumping measures. High among them was modified procedural and operational adjustments to enable faster and more effective protection from increasingly sophisticated unfair import tactics. While the federal government was not unsympathetic, neither did it fully embrace this cause. Officially, Ottawa held that the long-run solution required global agreement on capacity cuts and subsidy rules. In the meantime, it was willing to bring diplomatic influence to bear on states hosting proven unfair traders and to smooth certain administrative relations between the industry and the Canada Customs and Revenue Agency. To Canadian producers, however, the legal and economic rigidities of the CITT caused great frustration. The unwillingness of this quasi-judicial body to accept the policy shadings and political nuances that so marked its US counterpart—the ITC—was infuriating to Canadian steelmakers. In actual fact, it spoke to the political strength of steel-using manufacturers in the import trade equation.

Policy climates and discourses can change rapidly in the wake of events, and the Bush safeguard referral of June 2001 is a case in point. Well before the ITC reports were written, their watershed significance was clear. Whatever the outcomes, they promised to echo through the political corridors of NAFTA, the European Union steel councils, and the WTO apparatus. This recognition—that new levels of US safeguard protection were coming—was not lost in Ottawa. Officials in the departments of international trade, finance, customs and revenue, and industry were receptive to steel industry warnings on the imminent danger that offshore imports "diverted" from lost US contracts would end up in Canada instead. During the summer of 2001, a joint industry-government working group explored new forms of administrative cooperation, drawing attention to Canadian plans for responding to an import surge. This agenda covered improved gathering and reporting of statistics as well as new and more effective trigger mechanisms for import policing.

Canadian producers also had a crucial case to put before the section 201 hearings in Washington, arguing for the exemption of Canadian-sourced imports from the ITC review. In addition to the CSPA, testimony was offered by members of the Parliamentary Steel Caucus, an all-party grouping of 35 members of parliament who supported domestic producers in five provinces. Also important was the Canadian division of the USWA, whose ties to the larger US union were not insignificant in underlining the labour and trade dimensions of the continental steel market. Alliances forged in the heat of political battle are significant, and the coalition of interests backing Canada's ITC campaign of 2001-02 echoed the tripartite steel trade defence of the mid-1980s. In the end, the level of exemption from the ITC injury round, which covered hot- and cold-rolled and galvanized products but not bars or tubular goods, was viewed north of the border as a major victory for Canadian steelmakers.

The wider ramifications of the US safeguard proceeding, however, shifted the political focus back to Ottawa. In its findings, the ITC found serious injury to be caused by imports in 12 of the 33 product categories studied. Three months later, in March 2002, President Bush approved or exceeded the ITC tariff recommendations in a set of sweeping safeguard measures. This package can be compared in its political significance to the trigger price mechanism of the 1970s or the VRA regime of the 1980s. In reaction, European Union member states prepared challenges for the WTO. Other major steel exporters also reviewed their options, given this sudden and sharp closure of the US market, and commentators began to predict a "new steel trade war." Congressional politicians, big steel firms, and the unions came together in praise of Bush's bold strike.

For Canada, this posed several complications. The first was how to deal with the anticipated rise in steel imports diverted from former US markets. Mexico had responded quickly to the Bush announcement by imposing new steel tariffs to guard against just such a threat. Ottawa was under intense industry pressure for equivalent action. Following high-level meetings with Canadian steelmakers, the cabinet ministers for international trade (Pierre

Pettigrew), industry (Allan Rock), and financial institutions (John McCallum) announced on 21 March 2002 that measures were available should "critical circumstances" arise. This included the possibility of an emergency steel surtax should the level of imports start to surge. Second, the federal cabinet ordered the CITT to conduct its own steel safeguard inquiry and to report by the end of the summer of 2002 (Schmidt 2002). This marked the first such cabinet reference in almost a decade, signalling that steel imports were on the agenda at the highest ministerial level. However, the pivotal role of the CITT, given its mixed legacy of steel import injury findings, suggested that the policy outcome was not easily predictable.

THE INSTITUTIONAL SETTING

In Canada the mandate to defend industry against unfair import trading, both from dumping and for subsidies, is shared by two agencies. The Anti-Dumping and Countervail Directorate of the Canada Customs and Revenue Agency, formerly known as the Department of National Revenue, determines the validity of an unfair trade complaint and estimates the commercial cost of the violation. The Canadian International Trade Tribunal (CITT), a quasi-judicial body formerly known as the Anti-Dumping Tribunal, determines whether the activities in question have injured domestic producers and sets the final level of penalty duty. By the terms of the Special Import Measures Act (SIMA), Canada Customs must make a preliminary determination within 90 days and a final determination within an additional 90 days. If the preliminary is affirmative, the CITT then launches its own investigation of whether the evidence discloses a reasonable indication of material injury, retardation, or a threat to cause material injury. Once Canada Customs makes a positive final determination on the existence of dumping or subsidy, the CITT must complete a final injury inquiry within 120 days. In this process, the CITT follows a standard format. It solicits questionnaire data from the commercial interests that stand to be affected; prepares a staff report; conducts public hearings; and reports its determinations, its reasons, and its proposed remedies where injury has been found. Figure 3-4 captures the sequence of a SIMA dumping or subsidy action.

At present, CITT consists of seven appointed members, who form the investigating panels, supported by a staff of more than 80. The CITT also holds statutory responsibility for conducting safeguard inquiries. By contrast to a dumping or subsidy action, a safeguard inquiry is relatively rare. It is authorized under the WTO Agreement on Safeguards in cases where a sud-

Figure 3-4

Anti-Dumping and the Countervail Process

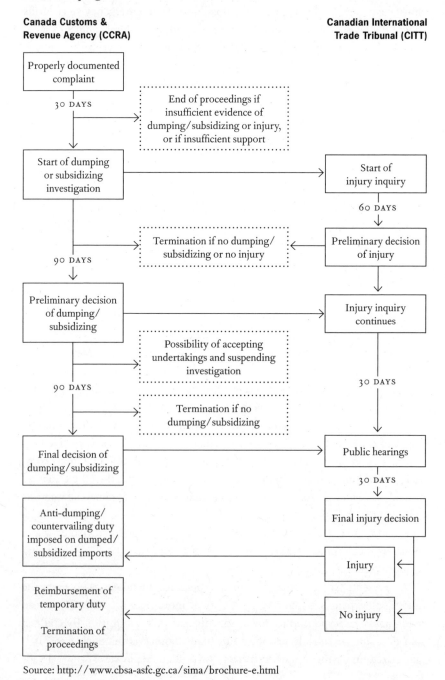

Source: http://www.cbsa-asfc.gc.ca/sima/brochure-e.html

den import surge either causes or threatens to cause serious injury to domestic industry. As already noted, the 21 March 2002 cabinet order for the Canadian steel safeguard investigation was the first such action in nearly a decade.

Procedurally, a safeguard inquiry differs from a dumping or subsidy inquiry by dispensing with the distinction between the preliminary and the final phase. Otherwise, it retains the notification, data-gathering, staff reporting, and public hearing steps, with the latter divided between an injury phase (June 2002 in this case) and a remedy phase (July 2002 in this case). It also involves the determination of injury and the recommendation of remedies, along with rulings on applications for exemptions from the remedies. In the steel case, the determinations were released on 4 July and the remedies on 19 August 2002.

Substantively, a safeguard inquiry differs from a dumping or subsidy inquiry in its scope and in its capacity to craft a balanced and compromised "package" of remedies. The scope is far broader, since a safeguard action is triggered by the likelihood of emergency disruption in an industry. Almost inevitably, the injury concern is sector-wide as distinct from product specific, as is normally the case concerning dumping and/or subsidies. This in turn opens the way for a safeguard ruling to reach variable product-by-product conclusions and to balance the overall outcomes if desired.

THE INJURY PARADIGM FOR SAFEGUARDS

While the procedure and purpose of a safeguard action are relatively transparent, the methods of calculation, standards of measurement, and modes of reasoning are far from obvious. In actual fact, a combination of trade case law and business accounting techniques provide the analytic tools for a safeguard proceeding. It is in their application, however, that the fine politics of steel trade relations are manifest.

To begin, consider the definition of the problem. The CITT was to "determine whether any of the goods specified is being imported into Canada from all sources *in such increased quantities* since the beginning of 1996, and under such conditions, as to be a *principal cause* of *serious injury* or *threat thereof* to domestic producers of *like or directly competitive goods*, on the basis of all relevant factors" (Canada 2002b, 9; author's emphasis added). Virtually none of these key concepts, standards, thresholds, or causal relationships are self-evident. Instead, the CITT drew upon its legislative base and procedural rules, together with trade treaty provisions and precedents from prior trade decisions and judicial reviews, both in Canada and abroad, to fashion a working methodology for the following points.

1) Which domestic steel goods are "like or directly competitive" with the subject import goods? This "classes of goods" question had to be addressed early in the process, as it effectively defined the categories of interest. The cabinet order used two different characterizations: three groupings of goods (flat products, long products, and tubular products) and nine specific types of goods (discrete plate, hot-rolled sheet and coil, cold-rolled sheet and coil, corrosion-resistant sheet and coil, hot-rolled bars, hot-rolled shapes and structurals, cold drawn and finished bars and rods, concrete reinforcing bars, and welded and tubular steel.) Sensing that the injury case could be made more effectively on a comprehensive scale, the Canadian steelmakers pushed for assessment on the "three groups" basis. On the other hand, foreign producers and importers pressed for assessment by the "nine types," which opened the way for selective injury findings, if any. The CITT opted for the latter, consistent with the categories routinely applied in dumping cases and to those followed by the US safeguard hearing of 2001.

2) What constitutes a significant "increased quantity" of importation of the subject goods since 1996? Drawing this time from its own regulations, the CITT stipulated a focus on the volume of import increases, both in absolute and relative terms. To this it added, from an earlier WTO appellate ruling, the requirements that such increases be "recent enough, sudden enough, sharp enough and significant enough to case serious injury" (Canada 2002b, 17).

3) Have the increased imports arisen as a result of unforeseen developments and the effect of obligations incurred under the GATT agreement? This provision of GATT Article XIX comes into play once the significant increase threshold (2) has been settled. While some steel importers suggested that the root problems were not of this "unforeseen" sort, thus nullifying the need for continuing the inquiry, the CITT ruled otherwise. It concluded that the key international steel market developments were indeed unforeseen and that Canada's treaty obligations had blocked it from responding to rising imports by, for example, imposing a general tariff. Thus, the requirements of Article XIX were satisfied.

4) Has the significant increase in imports imposed "serious injury" on domestic producers of the like or directly competitive goods? The CITT Act defines serious injury as "a significant overall impairment in the position of the domestic producers." The range of relevant factors in this calculation includes changes in level of production, employment, sales, market share, profits and losses, productivity, return on investment, capacity utilization, cash flow, inventories, wages, growth, ability to raise capital, and the impact of imports on prices (Canada 2002b, 19).

5) Is the significant increase in imports the "principal cause" of serious injury? Here the CITT Act defines principal cause as "an important cause that is no less important than any other cause of serious injury" (Canada 2002b, 20). It is also necessary that a causal link be shown between imports and injury. To satisfy this criterion, factors that could potentially cause injury, other than imports, must be examined. Should any factor other than imports prove more important, then the import factor cannot be considered a principal cause. Of all of the analytic calculations involved in a safeguard investigation, this is the most problematic. Not only is the exact parsing of causal factors very difficult in a technical sense, but it also carries the risks of achieving meaningless precision, given the complex interaction of factors such as steel supplies, prices, business cycles, product cycles, capacity utilization, and investment.

6) In the absence of increased imports serving as a principal cause of serious injury, are these imports a principal cause of "threat of serious injury"? The Act defines threat as "meaningful serious injury that, on the basis of facts and not merely of allegation, conjecture or remote possibility, is clearly imminent" (Canada 2002b, 21). In the steel safeguard case, the CITT assessed the state of the market in the opening half of 2002 and the prospects for changes to import levels.

7) In cases where imports are found to be a principal cause of injury or threat of injury, do imports from treaty partners in NAFTA, Israel, or Chile constitute a substantial share of total steel imports, and do these treaty imports contribute importantly to serious injury or threat of injury? Here a substantial share is defined as standing among the top five importers of the good over the most recent three-year period. If imports from a treaty partner do not match the overall surge of import levels, they will be deemed not to contribute to serious injury. Should a treaty partner's imports not be so deemed, it is exempted from the application of safeguard measures.

All of these steps figured in the CITT deliberations on steel safeguard measures. They were applied, in turn, to each of the nine designated product lines.

THE 2002 STEEL SAFEGUARD FINDINGS

The highly contested nature of the Canadian steel safeguard proceeding was reflected in both its scope and duration. More than 175 parties joined the inquiry, ranging from domestic producers to unions, importers, foreign producers, steel consumers, and government agents. It was, in the words of the final report, the most complex case ever conducted by the CITT, involving over 80,000 pages of documentation (Canada 2002b). Not surprisingly, legal and trade consultants played a central role in representing the parties. In fact, a breakdown of the parties according to counsel reveals that the top six consultants accounted for half of all parties while another 30 per cent of the intervenors represented themselves. Certain regional patterns are also evident, with particular consultants concentrating on European Union, German, US, Japanese, central European, and domestic Canadian firms, as well as Canadian importers.

Of the nine key product categories under review, CITT found that imports were the principal cause of injury in five product lines (discrete plate; cold-rolled sheet and coil; reinforcing bars; angles, shapes, and sections; and standard pipe). Imports were not, however, found to be the principal cause in another four product lines (hot-rolled sheet and coil, corrosion resistant sheet and coil, hot-rolled bars, and cold-drawn bars and rods).

In proposing remedies, CITT recommended tariff-rate quotas (TRQs) for four of the five principal cause products, rebars excepted. This stipulates an acceptable level of "non-injurious imports," above which a high tariff rate takes effect. In each case, a portion of the TRQ was reserved for US shipments. In the case of reinforcing bars, where there was no US involvement, a surtax was proposed instead. These remedy proposals drew significant criticism from Canadian steelmakers and the union. The use of TRQs suggests that import volumes are the key, whereas the burden of industry argument was that the downward pricing pressure from oversupply was the key to the disruption. Thus, the USWA maintained, Mexico's tariff-based response was more appropriate. Furthermore, there was much concern that by including US imports in the remedy package, the CITT jeopardized both the hard-won Canadian exemptions under the US safeguard action of 2001-02 and the future of the cross-border market. Finally there was intense frustration with the CITT's methodology in this case, which recognized the parts—the discrete steel product lines—without ever addressing the whole—the steel industry overall.

In some respects, the "five and four" findings were most widely publicized as the gatekeeping decisions that they were. Speaking for the domestic coalition of steelmakers, Stelco President Alfano found the findings extremely disappointing given both the urgency of the issue and the depth of the evidence. The chairman of the Parliamentary Steel Caucus, MP Tony Valeri of Hamilton, characterized the findings as "mixed" and suggested that they formed a basis for future effort. The Canadian director of the USWA, Lawrence McBrearty, was scathing in characterizing the report as a setback and "a shoddy piece of work that deserves to be ignored by the government"

(Erwin 2002). The USWA called on Ottawa to go further in providing remedies for a wider scale of injury and to move on reforming the legislative framework for fair import trading (McBrearty 2002).

THE FINDINGS FOR HOT-ROLLED SHEET AND COIL

To further illustrate the blend of technical and political considerations involved in safeguard proceedings, we focus here on one of the most hotly disputed findings, the CITT's treatment of hot-rolled sheet and coil steel (hereafter referred to as HR product). The domestic manufacturers in this case include the big three integrated firms along with IPSCO and Ipat-Sidbec. Their total 9.2m tonnes of HR product represents about half of the total steel output in the country. Much of this HR material is destined for further processing, either as feedstock within the companies of origin (57 per cent) or after merchant sale to end users or service centres (43 per cent). The import share of the HR market has fluctuated sharply over the period of safeguard study, from a low of 17 per cent in 1996 to highs of 36 and 35 per cent in 1998 and 2000 before settling to 19 per cent in 2001. All parties recognized the need for HR imports, though the domestic industry views the natural level at the lower end. Clearly, HR steel is a critical segment within the domestic industry, making the CITT findings highly consequential.

The CITT found that the rising HR import share was sufficient to meet the recent, sharp, sudden, and significant criteria. It also concluded that unforeseen developments, in the shape of the global overproduction linked to the Asian, Russian, and eastern European crises, had pushed low-cost imports into markets such as Canada. After examining the financial indicators, it was concluded that Canadian HR producers suffered serious injury, for which HR imports were a cause. However, the CITT concluded that other causes played the principal role. Key among these were "overall economic trends." During the years of strong demand in 1998-2000, imports were required to fill shortfalls in domestic supply. During this time, the CITT contended, even the Canadian producers imported some HR product to feed their downstream cold-rolled and galvanized mills. Then with the 2000 economic downturn, which was driven by declining automotive demand, prices weakened under pressure from domestic supplies. This was reinforced by the HR steelmakers' decisions, during the high demand years, to prioritize internal supplies over merchant supplies, thereby increasing the need for merchant imports. It was further exacerbated by the aggressive price-cutting by Algoma within the merchant market as it struggled to preserve market share.

In effect, the CITT found that the business strategies of Canadian HR producers, executed against the backdrop of rapidly changing domestic demand and continuing offshore oversupply, were the principal causes of injury. Critics of the CITT findings were far from persuaded by the analysis. They challenged the plausibility of disaggregating the causes of injury so starkly.

Can domestic demand be said to drive market outcomes when discounted imports make up one-sixth to one-third of total sales? Did domestic HR producers allocate their supply internally due to the prospect of higher return or due to the certainty of import pressure in the merchant market? If injury to an industry is the object of inquiry, is it advisable to fragment the analysis along product lines? Overall, it seems evident that the CITT viewed the recent import surges in the context of traditional steel strategies: to gear capacity to average rather than peak demand and to allow imports to fill the gaps. The domestic producers, on the other hand, argued that the new, ongoing reality of cheap offshore supplies is a transformational reality that infiltrates all strategic decisions. Finally, can a safeguard analysis that is, by definition, strictly retrospective to 1996-2001, accurately gauge a prospective "threat of injury" scenario that has yet to unfold?

Ultimately, the reaction that mattered most was that of the federal cabinet. Here, however, there was no quick response, as Ottawa wrestled with the dilemmas of whether to move beyond the CITT proposals, how to act within the strictures of the WTO and NAFTA, and how to handle the sensitive CITT findings on dumped US imports. The fact that the Bush administration was backpedalling throughout the summer of 2002, granting hundreds of exemptions from the March ruling in response to sharp price hikes and steel-user complaints of shortage, made it even more difficult to fashion a durable Canadian policy response. However, the combined results of a diluted US safeguard remedy and a relative resumption of US import flows lifted the urgency for a Canadian response. So too did the pressure from Canadian steel-users. Presumably, the 70-member auto caucus within the ruling Liberal Party helped remind ministers of the needs of steel-using manufacturers. So also, perhaps, did the August 2002 announcement by the Honda Motor Company that looming shortages of galvanized sheet steel had forced it to consider importing supplies by air from Japan (Brieger 2002). After much prevarication, the cabinet chose not to follow the CITT report. This decision will be discussed further in Chapter 6.

Overall, the safeguard case offers several insights into the politics of the steel trade. First of all, the pool of organized interests extends well beyond the domestic steel producers and unions. Every dumping, subsidy, or safeguard proceeding attracts a spectrum of foreign producers, import traders, and steel-user interests. Second, it is interesting to note the high degree of organizational coherence maintained by the Canadian steelmakers, whether they act through their industry association (the CSPA), through the less formal steel "coalition" grouping, or alongside organized labour through the Canadian Steel Trade and Employment Congress. There are few signs of the cleavage that divides US steelmakers into integrated and mini-mill camps. Third, it is clear that the CITT plays a critical role in arbitrating the fine differences among steel-sector interests at large. In its legislative mandate, its operating procedures, and its intellectual mindset, the CITT struggles to adjust the tensions and ambiguities associated with the policing of "fair trade." In another context, Rianne Mahon describes the role of semi-independent state agencies as one of managing unstable policy "compromises" that are destined to continual challenge (Mahon 1979). In effect, the federal state wants to maintain a vibrant domestic steelmaking industry while at the same time maintain an open continental and global trading system. It seeks to promote commercial competition as a path to efficiency and growth but seeks also to prevent unfair competition that established home industry cannot possibly meet.

Overall, then, the steel dumping and safeguards cases of 1997-2002 demonstrate the patterns of fine tactical politics between firms, product segments, and nations. It illustrates the ways in which the values embedded in rules and procedures can themselves serve as tactical levers and resources. It also illustrates the ways in which the terms of legal and economic discourse within the subfield of import trade policy shape the political combat. We have examined a series of ongoing campaigns aimed at confirming or denying the circumstances of dumping, unfair subsidy, material injury, short supply, and others. This can be expected to continue, with repeated probes and counterprobes by business rivals and occasional policy adjustments by state officials, until such time as new policy instruments are applied to the steel import problem.

Conclusion

By comparison to pulp and paper, the steel industry has a more restricted geographic footprint in Canada. Primary mills operate today in five provinces, though processing facilities and service centres are distributed more widely. Like pulp and paper, however, the steel sector exhibits extensive levels of backward and forward integration, significant sensitivities to raw material markets, and a pattern of supplier-customer relations that are intercorporate rather than retail. It is also interesting to note, once again, the profound impact of manufacturing technologies and innovations on the shifting commercial prospects of firms and subindustries.

Steel politics also appears to function more often through wholesale than retail channels of influence. This is not to deny the often profound social impacts of industrial action, largely contained, since the 1940s, through the labour relations regime, and mill facility closures, which triggered provincial state equity ownership in Nova Scotia, Quebec, and Ontario. Nor does it gainsay the roles of the Parliamentary Steel Caucus, the USWA, or even the air pollution lobby in shaping aspects of the policy agenda. However, it seems fair to say that the driving elements of steel politics are the industry blocs and sub-blocs whose fortunes are directly affected by heavy industry activity. This begins with the domestic steel producers but certainly does not end there. It extends to foreign steel producers and import-export merchants who broker shipments in Canada. Crucially, it also includes the steel-using industries in Canada, which possess a potent combination of market and policy power. This set of constituent interests is captured in Figure 3-5.

We have seen that a durable and mutually supportive compact was hammered out between industry and the central state following World War II. This policy-strategy consensus combined tax and tariff incentives with corporate capacity planning to generate dramatic improvements in productive efficiencies and cost competitiveness. Most importantly, the combined focus in high-capacity utilization and reliance on imports to serve residual demand avoided the more extreme supply-demand imbalances experienced by other steel markets at the extremes of their business cycles. As a result, the Canadian steel sector was in a strong performance position entering the 1980s. Critically, this policy served, at the same time, the needs of steel-producing and steel-consuming industries, resulting in a supportive political coalition of remarkable breadth and durability.

Even within this context, however, the contrasting fortunes of individual firms should warn against overgeneralization. Dofasco emerged as a technology leader and positioned itself effectively to serve the needs of the surging auto industry after 1965. Dosco experienced the increasing burden of spatial segmentation between Sydney and Montreal and ultimately collapsed under that weight. At the Nova Scotia end, its successor firm Sysco undertook an unusual and ultimately unavailing conversion from integrated to mini-mill operations. Algoma proved to be the weakest of the big three in weathering the stagflationary era after 1975, lurching awkwardly from takeover target to state enterprise to the bankruptcy courts. IPSCO arose from modest prairie beginnings to join the lead ranks of the multinational mini-mills. Stelco, of all

Figure 3-5

Key Elements of the Steel Policy Network

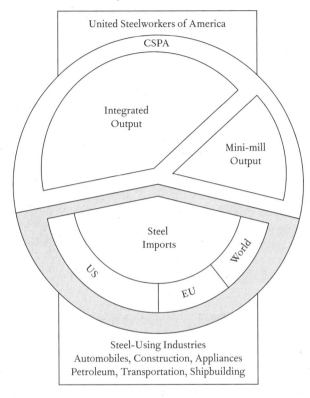

United Steelworkers of America

CSPA

Integrated
Output

Mini-mill
Output

Steel
Imports

US

World

EU

Steel-Using Industries
Automobiles, Construction, Appliances
Petroleum, Transportation, Shipbuilding

the Canadian firms, most closely resembles a US-style integrated steelmaker. These inter-firm differences continue to fuel the finely drawn politics of tactical advantage that erupt continually underneath the broader policy consensus. The Algoma-Sysco rail issue is a graphic reminder of such possibilities.

The struggle to reconcile the pursuit of enterprise interests with the pursuit of industry-wide interests is ongoing. In Canada, the steel sector has had to become more formally mobilized over the past generation and has gained significant experience in collective representation. The emergence of the Canadian Steel Producers Association and the business-labour Canadian Steel Trade and Employment Congress are two cases in point. More than anything else, this was driven by the emergence of continental trade as the new dominant policy plane. Since approximately 1980, the political imperative above all others has been the management of that 20 to 40 per cent of the steel market now persistently subject to international forces. The terrain for this encounter includes state regulatory authorities for dumping and safeguard policing—the Canada Customs and Revenue Agency and the US Department of Commerce, the CITT and ITC—the continental conditioning policy frameworks of the FTA and NAFTA, and the international forums of steel market adjustment such as the OECD and WTO.

New initiatives notwithstanding, the challenge of organizing a grand consensus in Canadian steel, equivalent to the one that drove the immediate postwar generation, has proven far more difficult. For steelmakers, the initial promise of the Canada-US continental trade negotiations was stillborn, particularly as the critical market access tools of common dumping and subsidy codes went unfulfilled. This left Canadian producers scrambling (along with other world suppliers) to escape and avoid damage by the highly politicized trade remedy regime in Washington. At the same time, the Canadian domestic steel market became increasingly vulnerable to foreign penetration, a problem for which dumping and countervail measures offer limited remedy at best. As we have seen, this intractable policy problem still awaits its political solution.

Key Terms and Relationships

Anti-Dumping and Countervail: Remedies against unfairly dumped or subsidized imports. Dumping is export sale at price below the cost of production. Countervail is a duty applied to balance the advantage gained by state-subsidized production.

Anti-combines: Policing of inter-firm conspiracy in restraint of trade or price-fixing.

Autopact: The 1965 sectoral trade agreement which restructured the Canada-US auto industry and increased Canadian-based production.

Bonusing: Nineteenth-century municipal grants to firms in return for plant location.

FTA/NAFTA: The 1989 and 1993 comprehensive trade agreements to eliminate tariffs and control non-tariff trade policies between Canada and the US in 1989, with Mexico added in 1993.

Demand Management: Application of Keynesian fiscal—revenue and expenditure—policy to stabilize consumer demand.

Industrial Relations: The legal framework for management-union relations including union recognition, contract bargaining, and strikes.

Legacy Costs: Continuing pension and health obligations of integrated steel firms for laid-off and retired workers.

Modernization Grants: Cash incentives to underwrite capital goods upgrade.

Multilateral Steel Agreement: Proposed sectoral trade deal among major steel-producing nations.

Procurement: Strategic state purchasing contracts with private suppliers (e.g., CNR rails).

State Enterprise: State ownership of industrial firms (e.g., Sysco and Sidbec after 1967).

Tariff: A duty applied to imported goods on entry, for protective or revenue purposes.

Tariff Board: State agency established in 1926 to regulate detailed application of Tariff Schedule in Canada.

Tax Expenditures: Tax-based incentives for capital spending (e.g., capital cost allowance).

Wartime Assets Disposal: Privatization of state-financed wartime steel plants after 1945.

CHAPTER FOUR

The Politics
of Air Transport

The Industry at the Millennium

The final and most momentous air policy decision of the 1990s was Ottawa's approval of Air Canada's takeover of its long-time rival, Canadian Airlines International (CAI). This brought to an end almost 70 years of business and political rivalry between two powerful corporate empires. It also re-established a degree of single-firm dominance not seen in air transport for half a century.

However, this was no simple case of "back to the future." Both the corporate players and the policy context had changed dramatically. Air Canada was no longer a state enterprise or Ottawa's favoured instrument for nation-building. By the year 2000 the airline was a shareholder-owned firm with an independent management. Far from enjoying privileged treatment from the governing party, Air Canada was regarded by many in Ottawa as a necessary evil. The western Canadian public identified strongly with CAI as its regional champion, and its capitulation to Air Canada was a bitter outcome, to be sure. In fact, a significant share of western customer loyalty shifted to the discount upstart, Westjet. Neither was it clear how the public interest could be protected from the monolithic survivor. Since the domestic air-sector deregulation of the 1980s, there has been no powerful agency to monitor and review matters of routes, fares, or service. Nor has an aggressive culture of business competition emerged to fill the gap. Some critics have even called for the abolition of ownership restrictions to enable US competitors to inject rivalry. The striking transition between these two air industry regimes will be explored as the chapter unfolds.

One of the central facts of today's scheduled air business is the past decade of unprofitability. With a few exceptions, the leading firms in both Canada and the US have stacked up steady losses since 1990. Paradoxically, however, there have been few notable business failures among the dominant firms. It has been said that, in airlines, conditions swing between the bad and the downright awful (Piggott 1998). In fact, the airline industry seems to be trapped in a circuit of overcapacity and marginal profits, with an occasional sprinkling of catastrophic events.

To understand the paradox it is necessary to consider a number of factors. One set of challenges arises from normal operating risks. The price of fuel is a key operating variable, which fluctuates sharply according to international markets. In 1999, aviation fuel was available for as little as US$0.26 per litre, when the price of crude oil ran at $12 per barrel. One year later it had risen to $0.38 per litre, an increase of more than 50 per cent. By mid-2001 the price of a barrel hit $28, and airlines began imposing fuel surcharges on tickets to counter their mounting losses.

Corporate debt loads proved to be even more punishing, and the pressures to borrow came from two directions. The airline industry is extremely capital intensive, and "fleet management" is a tactical challenge of the first order. The new aircraft models that have appeared in the past decade are far more fuel efficient than their predecessors. In addition, their wide body and dual aisle configurations offer improved passenger comfort and, thus, a significant competitive advantage. Consumers became more aware of fleet age, and leading operators felt relentless pressure to acquire new planes, even in money-losing years. The other factor pushing up debt levels was the need for borrowing to finance business mergers and acquisitions. It was the legacy of deals from the late 1980s that ultimately pushed CAI to its breaking point. After that, Air Canada was left to struggle under the massive costs of the 1999 CAI takeover. In 2001, business rival Canada 3000 ceased operations when its creditors moved in to seize the aircraft. In 2003, Air Canada filed for bankruptcy protection when its debt load became unserviceable.

Labour action is another constant in this highly unionized industry. Pilots, flight attendants, on-the-ground ticketing and baggage staff, and machinists have all mounted strikes in defence of wage packages or job tenure. A case in point was the threatened Air Canada pilots' strike in July 2000. With a two-year contract expiring in April, the pilots were more interested in job security than in Air Canada's offer of a 15 per cent wage hike over five years. The bigger battle involved the airline's plans to begin a merger of the separate union groups representing the erstwhile CAI and Air Canada staff, as well as those of their regional connectors. After a decade of relentless wage concessions at CAI, its workers were eager to join the higher scale, while Air Canada staff were determined to guard their seniority against the imminent layoffs and downsizing. Ultimately, the pilots' dispute was settled by mediation, but the tense labour relations persisted. In 2003, Air Canada's unions were willing to face a bankruptcy filing rather than absorb a unilateral job and wage cut.

Another set of challenges are more structurally based. Perhaps the most damaging factor of all is the cut-throat price competition among major carriers in both the US and Canada. Yield management refers to the art of filling planes with paying customers. It is true that the era of half-filled aircraft declined with deregulation. But even fully loaded planes yield only marginal returns when the unit price is held down by fare discounting. Put simply, too many airlines put too many planes in the sky, with the result that hardly anyone makes money.

Ultimately, one obvious solution is to reduce capacity through rationalization. This can take place either through business failures or through mergers and takeovers. Yet, as already noted, outright business failures have been rare in this sector. Two of the former US majors, Eastern and Pan American, failed in the 1980s. However, most of the several hundred failed airline firms, frequently cited by the deregulation lobby as a sign of competitive vigour, were of a lesser or secondary scale. In this the bankruptcy protection laws play a major role. Rather than exiting the industry, many of the troubled top firms opted in the 1990s for periods of court-supervised "Chapter 11" protection as an alternative to total collapse. This allowed them a temporary standstill and an opportunity to renegotiate union contracts or attract new creditors before emerging to limp along once again. In 2002, the damage spread to the very core of the US industry: to the triad of American, United, and Delta Airlines. A combination of post-09/11 security expenses, surging insurance charges, and slumping traffic levels prompted the industry leadership to predict a combined loss of US$7.5b on the year. In December the unthinkable happened, when industry giant United filed for bankruptcy protection. The following spring (2003), American Airlines narrowly avoided the same fate.

Mergers and takeover strategies have proved more common, though not always successful. The year 2000 saw a pair of significant merger initiatives, together with shifting international business alliances among major carriers. For example, British Airways and Royal Dutch KLM held merger talks that would have created the third largest global air carrier. This drew worried political attention from third-party states and at the European Union headquarters in Brussels, as well as in Washington. Ultimately, the deal foundered on Dutch worries about the loss of a national icon. Within months of this outcome, an earlier but unsuccessful plan to link British Airways and American Airlines was revived, with similar political anxieties and no immediate results.

In part because of the legal and political obstacles to formal corporate merger, many airlines have favoured the strategy of international business alliance. Independent companies, usually with complementary strengths, band together to coordinate flight schedules, booking codes, and frequent flyer programs in an effort to share the traffic. This avoids most of the state regulatory monitoring that proved fatal to so many .conventional merger schemes. In the summer of 2000, a new global alliance was announced. The SkyTeam partnership linked Delta Air Lines, Air France, Korean Air Lines, and Aeromexico in the third largest international carrier alliance, sharing 173 million passengers per year.

One possible solution to the profits crisis was seen by using bigger aircraft. In June 2000, Airbus Industrie revealed plans for its A3XX super-jumbo jet. This 555-seat, double-decker model is planned for delivery in 2005 and holds special appeal to long-haul and intercontinental carriers. It is only the most recent step in an aircraft manufacturing sweepstakes that pits the European-based Airbus against its American rival, Boeing.

Another solution lies in more liberalized access for carriers to international routes (Doganis 2001). While the US and Canada agreed to a significant expansion of cross-border service in 1995, the world air business remains heavily mercantile in nature, with states jealously guarding route and traffic controls. The year 2000 saw one potentially promising breakthrough: the first multilateral open-skies agreement between the US and the members of the Asia Pacific Economic Conference.

In Canada, 2000 was a year of adjustment following the "one big airline" marathon. For the last half of the preceding year, the future shape of the Canadian air sector had been up for grabs. The impending collapse of CAI, the Onex bid for a dual takeover and merger, the Air Canada counterattack, the Onex defeat, and Air Canada's mirror triumph were the prime coordinates of this drama. The costs, however, were considerable. Not only did Air Canada emerge with a far greater debt burden, but it also faced new legislative and regulatory measures in Ottawa. Life as a virtual monopolist is politically problematic. Air Canada was besieged by problems: from competitors alleging predatory behaviour, from record levels of consumer complaints, and from bewildering problems of first maintaining and then merging the two networks. This opened the way for a trenchant public debate on policy options and industry structures, the sort of debate that normally occurs just once in a generation. And still ahead were the 09/11 terrorist attacks of 2001, which threatened the viability of the air industry from new and unexpected directions.

Overview

In civil air transport, we see a contemporary service industry that has experienced remarkable growth over more than half a century but now finds itself mired in deep commercial turmoil. In part the history of the airline sector has been one of technical advances in aircraft performance, reflected in increased speed, reliability, capacity, and operating efficiency. It is also a history of a burgeoning postwar mass travel market in which holiday excursions and long-distance journeys ceased to be the privilege of the affluent few. The productivity gains from successive technological breakthroughs brought lower costs and lower fares, expanding travel markets, and new peaks of traffic. In Canada, the flagship airline doubled its passenger volume between 1944 and 1946 and again in 1949, 1953, 1958, and 1968 (Pooley 1996). While traffic volumes continue their explosive growth, the basic mode of passenger air transport has changed little since the 1970s, despite the arrival of new construction materials, control systems, and fuel efficiencies. Yet during this relatively short period, the economic and political shape of the industry has been transformed more fundamentally than in either of our other two cases. This began with the US Airline Deregulation Act of 1978, which overturned 40 years of state-regulated business practice (Brown 1987). The trend spread to Canada and Australia in the 1980s and to Europe in the 1990s. The compound consequences of the particularly Canadian style of liberalization continue to shape the scheduled air transport industry today.

As with all forms of transport, the commercial operations of the air industry depends on far more than the planes themselves. An often-forgotten component is the infrastructure that makes mass transport possible. Nowhere is this more acute than with airlines, which have the most exacting requirements for launch, navigation, landing, and service. As a business, the airline industry focuses on the economic indicators of demand (seats sold), return (revenue per seat kilometre), and efficiency (cost per revenue seat kilometre). Perhaps because of this, there is a tendency to ignore the tremendous infrastructural investments that until recently were the responsibility of public authorities (McGrath 1992). This includes runways and terminals, radar systems, air traffic controllers, and meteorological services. Infrastructural configurations and costs can have a decisive impact on the shape and performance of airline companies, conferring competitive advantage or disadvantage. Only when overburdened airports cause discernible disruptions of service, as occurred in the post-deregulation period, is this fact brought visibly home.

Figure 4-1

Long-Term Aviation Cycle

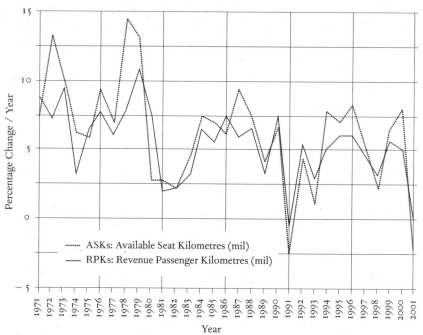

Source: Calculated from data by International Air Transport Association.

Yet this is only one aspect of state influence on the air industry. From the Depression until very recently, the air transport sector was subject to direct state regulation in most, if not all, of the Western capitalist economies. This was one of the central facts of the growth phase of the air transport product cycle. It directly affected the industry structure, ownership patterns, entrepreneurial outlook, investment parameters, technology adaptations, and the financial performance of firms. Explaining the political origins and ongoing dynamics of the regulatory regime reveals much about this industry. It is also important to note some striking variations in the level of maturity of geographic airline markets. While North America and Europe are furthest developed, the Asian market has entered a highly dynamic phase, and Africa is in an early stage. At the same time, the passenger air industry is subject to short-cycle fluctuations in traffic levels, as Figure 4-1 reveals.

Equally significant has been the prominence of state enterprise in the airline sector. Most modern states have looked to a national flag carrier as an expression of sovereignty and pride. Particularly in Europe and the Third World, the chosen instrument was frequently a state-owned company. By contrast, this question never arose in the US, where private capital was hegemonic. Curiously in light of its subsequent dominance, the Canadian crown carrier was established only when private capital declined a state invitation for partnership. This window of private investment opportunity proved to be narrow, as an elaborate regulatory framework was imposed on the growing Canadian air industry beginning in 1938, shaping its structure for more than 40 years. This model, which Canada shared with the United Kingdom and Australia, was "a mixture of public and private enterprise, regulated by a government authority" (Corbett 1965, 83).

Yet since the late 1970s, this trend of state policy has been reversed. Beginning with the US initiative of 1978, many national airline industries have been deregulated by degrees, Canada among them. This triggered an unprecedented transformation of domestic air markets, manifest in carrier competition, flagship privatization, traffic growth, and takeovers and bankruptcies. Together they made air transport one of the most turbulent of modern service industries. Although the traditional economic controls were removed, there are reasons to question the overall extent of industry deregulation. Events in both Canada and the US suggest that quite apart from continuing responsibilities for infrastructure and safety, the state is still faced with a complex range of choices and decisions. Many of these are the product of the deregulated environment (Dempsey and Goetz 1992). In fact, it could be argued that today's air policy conundrum is that the state is still expected to play a directive role but now lacks the legal and financial policy capability it once possessed for strong, coherent responses.

These remarks are increasingly relevant to the international air industry as well. That market is neither regulated in the traditional domestic manner nor open to all entrants. Instead, an antiquated form of mercantile diplomacy lingers on, with most states committed first and foremost to the advancement of their national carriers, whose prospects are strongly determined by reciprocal negotiations of route and landing rights. Yet many states and airlines are pursuing wider strategies on an integrated global scale. One example is the "free trade" movement for air services, while another is the trend to global alliances among leading air carriers.

The deregulatory initiative coincided with several other notable developments. One was the explosion of finance capitalism in the 1980s. This meant that considerable capital was available in the newly liberalized setting to underwrite new ventures or takeovers and mergers among existing carriers. It also elevated financial managers to new heights of influence within the industry, with predictable effects on entrepreneurial style. Second was the volatility of both major business cycles during this era. In the US, the 1982 recession, the most severe since the Depression of the 1930s, occurred early in the air industry restructuring process. By the time of the next trough in 1990-93, the entire global airline industry was caught up in an unprecedented profit squeeze from which it was slow to recover. In this light, it is hardly surprising that severe structural pressures have come to bear on the high wage, unionized workforce of the airline industry. In contrast to the case of the steel industry, the principal source of competitive pressure has been domestic rather than import-based. Yet, like steel, the catalysts often have been new or reorganized domestic firms with non-union workforces and sunbelt business outlooks.

This chapter continues with a brief history of the air industry and state air policy in Canada, which highlights some of the turning points and transitions in sector performance on the road to the present. Attention then turns to the structure of the modern industry, before and after the deregulatory watershed. Ironically, in light of the model role of the US experience, Canada experienced an airline duopoly — two-firm dominance — far tighter than the US air oligopoly. In both cases, a limited number of business strategies can be discerned in the new market. For corporate profiles, we examine the strategies of the two dominant Canadian companies of the 1990s: the original flagship carrier, Air Canada and its long-time competitor, Canadian Pacific Airlines (CPA; later CAI). The contrasting internal politics of these two firms explains much about their business fortunes. Two intriguing policy case studies round out this chapter. The first involves international business, through Canada's efforts to forge a new bilateral air agreement with the US. Achieved in the mid-1990s, this was based on the apparently liberalized principles of "open skies." By opening a new trans-border service segment of huge potential, it altered the air business environment in a dramatic way, at least for firms that were positioned to take advantage. The second case turns to domestic air politics and the most momentous development in Canadian air policy. It analyzes the federal government's handling of the CAI crisis of 1999 and the Onex-Air Canada battle for control of the one big airline. This resulted in the virtual monopoly whose political complications will continue to define the air policy agenda for the foreseeable future.

Historical Development

The pioneers of air travel achieved their decisive breakthroughs soon after the opening of the twentieth century. In 1903 the Wright Brothers made their first famous flight at Kitty Hawk, North Carolina, where they remained airborne for 59 seconds. The following year they flew a continuous 24 miles. In 1908 Howard McCurdy recorded the first flight in the British Empire, at Baddeck, Nova Scotia, while Louis Bleriot crossed the English Channel from Calais to Dover in 1909 to claim the first international flight. The single-seat aircraft of pioneer days was adapted for military use in World War I, where it proved its worth in both reconnaissance and combat. Soon after, Alcock and Brown completed the first trans-Atlantic crossing in 1919.

By then, the aviation engineering race was well underway (Baldwin 1975). First came the fully enclosed fuselage and the possibility of heated and pressurized cabins. The next generation of multi-engine propeller aircraft, exemplified by the Dakota DC-3, could ferry freight and passengers over longer distances at unprecedented speeds. Again, this had a military counterpart in the fighters and bombers that played such pivotal roles in World War II. In the post-1945 period, it was evident that jet propulsion could open yet another generation of air transport. It offered a bridge to long-distance travel and high-volume passenger traffic that culminated in the "jumbo jets" of the late 1960s such as the Boeing 747 and DC-10. By the 1990s, both Boeing and Airbus were planning their next generation of aircraft. The two most frequently discussed designs were the super-jumbo, which would double maximum capacity to as many as 800 passengers, and the new supersonic aircraft, flying conventional loads at speeds beyond Mach 1.

The interwar period also saw the expansion of commercial air transport outlets. In southern Canada, many of the early aerodromes were operated by private flying clubs. Then during the 1920s, as the possibilities of air mail were recognized, the federal government let postal carriage contracts to airline operators. This was a politically sensitive business, since the long-term guaranteed mail traffic often conferred decisive revenue advantages on the contractors over their competitors. In the US a particularly aggressive postmaster general, Walter Brown, used these powers to promote airline mergers in 1929 and faced anti-trust hearings five years later (Sampson 1984).

The Canadian northland offered another avenue for commercial flying, with small "bush airlines" serving hundreds of isolated communities and resource towns (Main 1967). Many of the legendary names in the industry began as bush pilots, flying float-planes, ski-planes, and freighters on the frontier. It was also in this area that Canadian aircraft manufacturers such as deHavilland developed the comparative advantage that opened global markets for models known as Beavers, Otters, Caribous, and Buffalos.

As the industry assumed ever-greater commercial importance, state jurisdiction over air transport became an inevitable issue within Canada's federal framework. This question was largely settled by the terms of the *Aeronautics* Reference case, determined by the Judicial Committee of the Privy Council in 1932. By locating aeronautics under the dominion's treaty-making power (section 132 of the constitution), the court cleared the way for central government jurisdiction. Direct state regulation of airlines in Canada began in 1936, when the Board of Transport Commissioners, originally established to deal with railways, had its authority extended to aviation (Ashley 1963). One year later, Trans-Canada Airlines (TCA; now Air Canada) was established as a central state-owned enterprise. For regulatory purposes, TCA was treated as a special case. It was controlled by the government of Canada by means of a contract with the federal minister of transport. For all other carriers, the Board of Transport held significant licensing powers for the approval of fares, schedules, route patterns, and types of aircraft. Toward the end of World War II, a separate Air Transport Board came into being, as a quasi-independent agency reporting to the minister of transport (Harris 1978).

An early, if unsuccessful, attempt to reverse this process and partially "deregulate" the airlines occurred in the late 1950s. The Diefenbaker Conservative government sought to curb the powers of the Air Transport Board. Its aim was to increase the level of competition among carriers and to open opportunities for CPA in particular. Opposed by most of the airline industry, this initiative was abandoned once the Liberal Party returned to power in 1963.

Three years later, Ottawa announced its new policy for "regional" air carriers (Baldwin 1975). Further refined in 1969, this framework shaped the modern industry profoundly, with a legacy that can still be seen today. The Regional Air Policy distinguished several "tiers" of air carriers, with little overlap between them. The two transcontinental carriers, Air Canada and CPA, formed the first tier. Below them stood five regional carriers: Pacific Western (PWA), Transair, Nordair, Quebecair, and Eastern Provincial Airways. Licensed for jet aircraft, each was given access to a segment of the far north, as well as to a core regional market with hubs and feeder routes. In 1969 each of the regionals was allotted an exclusive territory: British Columbia and Alberta for PWA, Saskatchewan and Manitoba for Transair, Ontario and western Quebec for Nordair, the rest of Quebec for Quebecair, and the Atlantic provinces for Eastern Provincial Airways. The third tier consisted of smaller scheduled and charter carriers with propeller-driven aircraft.

The 1966 policy was in large part a confirmation of the prevailing structure of the industry. However it also gave the Air Transport Board the authority to make modifications in the future. In effect, the minister of transport used the 1966 statement to provide direction to the regulatory agency, while at the same time clarifying government intentions to the industry. In 1967 the Air Transport Board lost its independent identity when it was consolidated into the newly formed Canadian Transport Commission, the CTC (Janisch 1979). One of the CTC's four modal committees, the new Air Transport Committee, assumed the air-sector mandate. With its domestic policy framework established, Ottawa turned next to the international scene. In 1973 it announced a policy for overseas carriers based in Canada. As with the regional policy, it assigned separate global areas of service to the two carriers, Air Canada and CPA.

The late 1970s saw several new policy initiatives in the domestic field (Langford and Huffman 1988). One of these was designed to reduce Air Canada's special status. The Air Canada Act of 1977 terminated the firm's contractual relationship with the minister of transport. It also transformed Air Canada's debt funding into common shares, held in state hands, and amended the Aeronautics Act to bring Air Canada under the same Air Transport Committee authority as the private carriers. At the same time, the route licensing rules were loosened for CPA, affording it a larger share of the transcontinental market. By 1979 CPA held a 45 per cent share of this market, compared to only 12 per cent in 1967. The area of air charter regulations, traditionally quite tightly controlled in order to protect the scheduled operators, was also reviewed. As a result, the terms for charter bookings became far more flexible on both domestic and international routes. Shorter booking periods, seat sales, and the mixing of scheduled and charter passengers on common flights, were all introduced as a result. Stanbury and Thompson (1982) described these developments as a compromise, giving the appearance of deregulation without the substance. While this appears to have been satisfactory to many air carrier interests, more powerful pressures for change were already building.

Still wedded to the three-tier policy, the Ministry of Transport released some modest reform proposals that were referred to a parliamentary committee, the Standing Committee on Transport, for review (Canada 1982). The results were predictable. Airline firms and their unionized workers stressed the virtues of stability, service, and safety, while consumers decried the high cost of air travel and looked south of the border for models of relief. By this time, of course, the US was well into its liberalization experiment, having enacted the Airline Deregulation Act in 1978. In its comprehensive 1981 report on *Reforming Regulation*, the Economic Council of Canada made a strong case for similar air sector measures in Canada (Canada 1981).

It was only after the appointment of a new transport minister, Lloyd Axworthy, that the paradigm began to shift. An experienced cabinet hand,

Axworthy brought with him a large ministerial staff group that was experienced in bureaucratic politics (Bakvis 1991). One of the minister's first
symbolic steps in the air portfolio was to cancel the complimentary airline
passes held by certain Ministry of Transport and Air Transport Committee
staff. He also directed the latter to launch a review of fares policy, while a
task force was struck to consider ways of adapting US-style deregulation to
Canada. Speaking as much to the Air Transport Committee as to the airlines
themselves, Axworthy announced that, henceforth, his office would look very
favourably on any appeals by air carriers where their applications for fare or
route competition had been denied (Axworthy 1984). This constituted the
de facto deregulation of scheduled service and opened a turbulent period of
restructuring (Button 1989).

It remained, however, for the newly elected Mulroney government to
formalize this new policy in law. A White Paper outlining the new liberal
philosophy was released early in the term (Canada 1985). The new National
Transportation Act, which took effect early in 1988, eliminated the traditional regulatory controls over routes, fares, and services, with their inherent
bias toward protecting existing operators. Instead, the renamed National
Transportation Agency was directed to license all applicants that were "fit,
willing, and able." In both countries, the airline industry was entering un-
charted territory (Lazar 1984), discussed more fully in the sections below.

Somewhat surprisingly, the rash of mergers and takeovers between 1984-88
never figured as a major question of public policy. Instead, it was treated as
a natural, if unpredictable, structural shakeout (Reschenthaler and Stanbury
1983). Yet these developments proved to be critical to the evolving shape of
air transport in Canada. In contrast to the short decisive deregulatory stroke
in the US in 1978, it took five years to finalize the competitive air framework
in Canada. While the CTC remained in place until December 1987, the leading
carriers certainly took advantage of the declining conviction of regulatory oversight following 1984. Consequently, the first phase of Canadian deregulation
was typified not by new entrants and enhanced competition, as in the US, but by
mergers, takeovers, and increased concentration among existing carriers.

Ironically, these were the years when the long-deferred revision of
Canadian competition policy was also underway. A decade of fitful reform
culminated in 1986 with the passage of a new Competition Act. In fact, during the final years of direct air industry regulation, the "regulated conduct"
exemption in the law raised doubts about whether the competition law could
even be applied in transportation. (This sprang from the Supreme Court of
Canada's 1982 ruling in the case of the *BC Law Society v. Jabour.*) In response,

Mulroney's new air policy regime established what was in effect a dual authority. On the one hand, the new National Transportation Act gave the National Transportation Agency the authority to review mergers and take-overs exceeding $10m in assets or gross sales for their impact on the public interest. This power was soon tested in the 1988 PWA takeover of Wardair, a transaction which was approved by the new-look National Transportation Agency. In 1993, it was applied also to American Airlines' proposed 25 per cent equity purchase in CAI, which was also approved. On the other hand, several sections of the 1986 Competition Act, including those covering preda-tory pricing and the abuse of a dominant position, held potential relevance to the emerging duopoly in airline transport. These applied particularly to the new barriers to entry associated with computer reservation systems (CRSs), inter-company code-sharing arrangements, frequent flyer programs, and airport slot allocation.

Arguments for the necessity of effective competition enforcement were raised forcefully during the deregulation policy debates (Gillen, *et al.* 1988a; 1988b), but to little effect on the outcome. In truth, a vigorous competition policy was foreign to both the intellectual perspectives and policy networks of Canadian air transport. Traditionally the air industry policy regime centred around the Department of Transport and its regulatory agent. It was here that the gradual liberalization of the early 1980s was charted and, despite many subsequent changes the air carriers remained a nimble clientele who had ne-gotiated their way through the regulatory hearings, the Axworthy initiative, the White Paper, the Parliamentary Committee hearings, and the passage of Bill C-18 (the National Transportation Act). The transport authorities and the airline firms shared a working perspective on the way the world worked.

By contrast, the competition policy regime developed quite apart from the air sector. Its bureaucratic base lay in the Department of Consumer and Corporate Affairs, itself a strange amalgam of clientele interests that had little traditional bearing on transportation industries. Within its Bureau of Competition Policy, the intellectual paradigm was drawn from industrial organization, and the policy problem centred upon the promotion of competi-tive markets. This was both foreign and threatening to air industry interests. Moreover, the Bureau had shown a long-standing concern with air transport policies and raised hard questions about rigid markets, excessive fares, and the need to defend consumer interests (Jordan 1983). Given the starkly inef-fective history of anti-combines policy in Canada, and its lack of application to service industries, the air network may have felt little threat in the past. However, a rejuvenated and expanded competition regime was a different matter altogether. In this context it is pertinent to note the verdict of Oum, Stanbury, and Trethaway that "Canada did not do a good job in shifting the mode of social control over the domestic airline industry from direct regulation to competition policy" (1991, 19).

In this looming policy vacuum, the defective policy design of air deregulation was also starkly revealed. As discussed earlier, the mandate tensions between the National Transportation Agency and the Competition Tribunal were never satisfactorily resolved. The resulting gaps and overlaps contributed to the confusion and opened a seemingly endless potential for litigation. It created a deregulated world of "quasi-politics," where conflicting interests battled before quasi-judicial agencies and courts of law, with no strong strategic sectoral direction (Campbell and Pal 1994, 86-87). As will be seen below, this sowed important seeds for the CAI-Onex-Air Canada battle of 1999.

Structure

There is a general perception that the air transport industry has evolved historically from commercial, though state-subsidized, origins, through a period of extensive state regulation, to the present liberalized market stage where direct state involvement is confined to infrastructure and safety matters. While this certainly applies to the scheduled passenger segment, it is not true of the industry as a whole. A comprehensive portrait of the air industry would include the following segments: air service for hire or lease, privately owned and operated air fleets, courier and parcel express, cargo and freight, scheduled passenger service, and charter passenger service. In strict terms, only the final three fit the paradigm sketched above. In this chapter, we focus on the passenger business, which may in turn be subdivided several ways. One familiar distinction is between scheduled and non-scheduled (or charter) activity. The former offers publicly accessible service on fixed routes and times, while the latter contracts with private clients for restricted services subject to special conditions. Traditionally, the scheduled carriers accounted for the far greater proportion of traffic but operated at lower overall capacity than did charter carriers, with the result that scheduled fares were significantly higher. Regulators took pains to prevent charter operators from eroding the core travel market on which the scheduled trade depended.

The scheduled and charter segments are sufficiently different to qualify as separate businesses as defined by Magaziner and Reich (1982). The global peak for the charter segment came in 1972 when it exceeded 30 per cent of all traffic. Although the distinction was once watertight, it began to erode in the

1970s when scheduled operators were permitted to offer more flexible booking terms and fares, and charter carriers were allowed to take individual as well as exclusive group bookings. By selling a portion of seat capacity on long-term advance booking with no rights of exchange—the so-called advance booking charters—scheduled carriers boosted their capacity utilization, while charter carriers gained access to a far wider travel market by doing much the same thing. It was in this period that the Canadian company Wardair achieved international prominence as a trans-Atlantic charter carrier. Although the US charter trade collapsed following deregulation in 1978, Wardair remained profitable until its 1987 entry into competitive scheduled service. When deregulation began in Canada in 1983, the charter segment became extremely volatile, with extensive corporate turnover. In the 1990s several charter firms like Air Transat, Royal, and Canada 3000 combined charter with scheduled services.

This leads to another key distinction, involving domestic and international transport. A nation state can exercise sovereignty in its own air space and shape the domestic air industry accordingly. However trans-border flights are subject to international agreement between the states involved. Since 1945 a complex international regulatory regime has evolved to govern traffic patterns, the allocation of route rights to carriers, and permissible fare levels. This will be taken up in more detail in a later section, but it is important to note this spatial segmentation within the passenger business, since it affects the business strategies of most airline firms. The two segments are often combined, however, for purposes of aggregate global airline rankings. As Table 4-1 reveals, the largest carriers based on revenue are American, European, and Japanese. In the 1990s, the Canadian duopoly of Air Canada and CAI occupied positions in the low and high twenties, respectively. More recently, the consolidated Air Canada ranks just outside the top dozen global air carriers.

It should be evident from these preliminary comments that air transport has been by far the most politically managed market of the three industries that we are studying. By applying the regulatory powers conferred by statute, state agencies in the US and Canada traditionally determined the number of air carriers, their respective market shares, and the overall seat market. Invoking the highly discretionary standard of "public convenience and necessity," the regulators followed a cautious approach to licensing and fare approval, which was aimed at assuring existing carriers sufficient returns to guarantee reliable, safe, and profitable operations. In cases such as Canada, where a state-owned flag carrier was designated the lead company by formal policy, the private carriers were often further confined to junior roles and subject to different rules.

Table 4–1

World Top 20 Scheduled Airline Groups, 2002 (Ranked by Revenue Passenger Miles)

Rank	Group/Airline	Revenue Passenger Miles (mil)	Total Revs. $US (mil)	Total Revs. 2002/2001	Net Income $US (mil)
1	American Airlines	121,646.20	$ 15,730.7	+ 2.3 %	$ − 3,511.0
2	United Airlines	109,378.80	$ 12,719.1	− 14.3 %	$ − 3,212.0
3	Delta Air Lines	95,207.90	$ 11,720.8	− 8.2 %	$ − 1,272.0
4	Northwest Airlines	72,002.20	$ 8,988.4	− 4.8 %	$ − 798.0
5	British Airways	61,594.00	$ 11,552.3	− 4.6 %	$ + 108.2
6	Air France[a]	61,232.30	$ 11,981.9	+ 28.6 %	$ + 113.3
7	Lufthansa Airlines	51,188.40	$ 11,431.0	+ 21.9 %	$ + 681.2
8	Continental Airlines	56,830.60	$ 6,883.3	− 8.7 %	$ − 451.0
9	Japan Airlines[b]	51,697.10	$ 13,120.1	+ 36.9 %	$ + 73.1
10	Singapore Airlines	46,089.60	$ 5,875.6	+ 9.8 %	$ + 625.4
11	Southwest Airlines	45,396.10	$ 5,508.4	− 0.5 %	$ + 241.0
12	Quantas Airways	45,293.30	$ 5,897.2	+ 7.8 %	$ + 223.6
13	Air Canada	42,887.60	$ 6,220.6	+ 2.9 %	$ − 524.2
14	US Airways	40,024.00	$ 5,632.0	− 21.1 %	$ − 1,646.0
15	KLM Royal Dutch[a]	36,595.80	$ 6,124.6	+ 20.8 %	$ − 392.9
16	All Nippon Airways	33,691.90	$ 10,132.6	+ 5.3 %	$ − 235.0
17	Cathay Pacific	30,454.90	$ 4,242.3	+ 8.7 %	$ + 513.2
18	Thai Airways Int'l	30,036.20	$ 2,998.8	+ 1.3 %	$ + 237.2
19	Korean Air	25,749.90	$ 5,206.3	+ 17.8 %	$ + 93.2
20	Liberia Airlines	25,116.50	$ 4,438.3	+ 4.7 %	$ + 93.2

[a] In 2003, Air France and KLM announced a merger plan.
[b] In October 2002, Japan Airlines began integrating newly acquired Japan Air Systems.

Notes: Total Revenues exclude incidental revenues; Net Profit is after taxes.
Source: Aviation Week and Space Technology, *2004 Aerospace Sourcebook*, 19 January 2004, 331–66.

There were many business achievements of the regulated airline business. However, the management outlook enshrined by this regime was cautious, rather unsophisticated (in comparative business terms), and heavily dependent on technical advances to enlarge core markets and to create new ones. The rise of the budget tourist market in the 1970s is a case in point. Gialloreto characterizes the managerial outlook this way:

> growth was a slow painful process and … any product ideas that
> deviated from the norm in terms of value (price for service ratio)
> were unlikely to be accepted … pleasing the regulators was more
> important than pleasing the customers. (Gialloreto 1988, 13)

Certainly this could no longer apply to the last 20 years, as state regulation has declined and new forms of competition have emerged. For example, the relaxation of route licensing, coupled with the ready availability of used aircraft for purchase or lease, went some distance to eliminate the barriers to new entrants, particularly in the early years. Yet the Canadian market failed to attract the same rush of new competitors that was seen in the US during the 1980s. Instead, the first five years of liberalized competition in Canada was dominated by mergers and takeovers among existing players. Apart from their aircraft and workforces, airlines possess two valued assets: their route structures (and client loyalty) and their airport "slots" (dedicated loading, take-off, and landing facilities). Both of these became strategic commercial commodities.

As far as pricing was concerned, the immediate expectation from deregulation, as well as the central strand in its rationale, was lower prices. In fact prices changed in both directions. While fares fell on the main trunk routes, making it difficult for any carrier to make a profit there, fares rose significantly on many secondary routes that lacked competition. As a result, consumer organizations have tended to view deregulation as a mixed blessing. There can be no question that fare structures became much more volatile. On the trunk routes, price wars among major carriers became regular fixtures during slack seasons, such as mid-winter. In addition, new entrants tended to rely on fare-cutting as their principal tool for carving out a market share. As a result, passengers became permanently price sensitive.

Another equally important factor was the proliferation in types of fare according to booking terms and class of service. Here firms adopted complex and constantly changing classifications. Where previously carriers could distribute quarterly or biannual directories to travel agents, the new system could only be delivered by on-line CRSs. This allowed agents to scan for optimal tickets among the multitude of available fares and carriers. Equally it enabled the airlines to perfect "yield management systems" by which they pursued higher capacity by continually adjusting seat space by price and by service class in response to demand. Firms could announce seat sales, discount particular services, and match the pricing moves of their opponents. On the one hand, this brought customers and travel agents very close to a condition of perfect information. On the other, it opened an avenue for cartelized behaviour through price leadership and parallel pricing. The term "silent conversations" refers to leading companies' use of their CRS systems to warn off competitors who threaten discounted prices (Avmark 1993). In fact, the US Justice Department brought a suit against eight leading carriers in 1991, alleging price fixing through computerized fare exchanges over the preceding three-year period.

Another dimension of the new airline competition lay in the promotion of rival CRS networks. In the late 1980s there were no less than five US systems, one Canadian, and two European. This battleground was two dimensional. Each CRS was sponsored by a single airline or consortium, which not only linked its own ticket and reservation offices but sought maximum penetration of the private travel agent industry (Humphreys 1994). The fabled battle between American Airlines' Sabre system and United's Apollo raged for most of the 1980s before Sabre emerged in control of a dominant share of the booking sector. Secondly, each booking system sought to recruit additional carriers, both by commercial deals and takeovers, to enlarge its data base and enhance its value. As the CRS rivals managed to sign up exclusive clients, the information flow to consumers became fragmented, and new corporate alliances were formalized through agreements on "code-sharing." The role of CRS systems as high value-added services is well recognized in the industry. Significantly, one of the key conditions and most intractable difficulties in American Airlines' 1992 rescue offer for CAI was insistence that the latter switch from the Gemini system in Canada to the Sabre system. This was expected to be more profitable to American than its share of earnings from CAI operations!

An important aspect of non-price competition can be seen in the emergence of "frequent flyer" programs. First appearing in Canada in 1984, these schemes constitute a powerful form of differentiation for an otherwise quite homogeneous product (Williams 1995). Airline firms seek to capture the exclusive business of regular customers by offering "free" tickets or service upgrades to clients who amass high volumes of travel miles. It is a form of marketing tool that could only be launched in a commercial market. Significantly, it is most effective in the hands of a major firm with a broad enough route structure to accommodate most of the needs of card-holders.

These were some of the hallmarks of the dynamic new environment that firms faced after 1980. Louis Gialloreto (1988) pointed out three generic business strategies that emerged in the US as the "golden rules" of the regulatory era were abandoned. Many of the industry leaders tried to continue as *high-cost, full-service carriers*. This was done in the context of the new hub-and-spoke networks and of the code-sharing connections with commuter airlines that had become the norm by 1985. These carriers were particularly vulnerable to industry recessions, which hit in 1982-83, 1991-92 and 2000-01. While three of the US leaders were able to survive and at times prosper under this strategy, many former pillars including Eastern, Pan Am, and Continental could not. Both Air Canada and CAI fit into this category.

A second approach lay in *low-cost, low-service operations*. Flexible workforces, lower wage rates, and tight debt control permitted effective price competition in basic service categories. The pioneer in this category was Pacific Southwest Airlines. In 1959 it began service against three national airlines on the Los Angeles-San Francisco route. As an intra-state operator, Pacific Southwest fell under the jurisdiction of the California Public Utilities Commission, which had no power to limit entry and, as a consequence, approved virtually all fares filed by carriers (Kahn 1988). By 1962 Pacific Southwest's market share on the route had grown from 13 to 43 per cent. Today, Southwest Airlines remains the exemplar of the low-cost, high-return carrier, and has grown to rank among the world's top 15 carriers. In Canada, this proved an effective entry strategy for the Calgary-based Westjet in 1996 and the Halifax-based Canjet several years later.

Gialloreto's third strategy was a hybrid based on *low-cost, medium-differentiated service*. This was an option for firms that began as type one or two carriers but were able either to cut costs, by renegotiating labour contracts or improving hub structures, or to shift from basic to enhanced service, including business class, with higher revenue yields. Over time, however, firms attempting this transition proved unstable and short-lived, and the relevance of this third option can be questioned.

Today, the passenger air industry has evolved into a two-segment business in most parts of the world, though the balance between the two is still conditioned by national market and political forces. Most countries are still served by full-service network carriers (FSNCs), which follow the standard business model of the postwar period (Trethaway 2003). They sustained high-cost operating structures through effective price discrimination. High fares for short notice and quick return travel were combined with enhanced service levels, while lower fares were available for excursion bookings, particularly with the "Saturday stay-over" condition. At the same time, the low-cost carrier (LCC) has also become a fixture in the major travel markets of Europe, North America, and Asia. Its advantages extend well beyond low operating costs. LCCs succeeded in breaking the traditional airline business mould by abandoning price discrimination and offering a single basic travel class together with the flexibility of one-way, short notice bookings. The gaps between FSNC and LCC fares, on comparable routes, are often on a 4:1 or 3:1 ratio. What the LCC surrenders in terms of unit fare, it more than recovers in higher load factors, measured by seat occupancy. Since 1997, for example, Westjet's annual load factor has remained above 70 per cent, whereas its break-even load factor level lay between 60 and 65 per cent.

Michael Trethaway (2003) offers an intriguing prognosis for these two segments in the passenger air sector. He contends that the decisive revenue-cost advantages enjoyed by LCCs will only grow with time. The best available response by the FSNCs will come from one-time restructuring shifts in response to deep losses or reorganization in bankruptcy. LCCs, on the other hand, can progressively reduce fares as traffic increases, thereby shifting the bar for future efficiencies. With LCCs accounting for more than one-quarter of domestic passengers in North America in 2003, the forecasts for future domestic passenger traffic run as high as 50 per cent. FSNCs may retain an added market share through affiliated, "quasi-low cost" operators like Air Canada's Tango (long haul) and Zip (short haul).

However, Trethaway contends that FSNCs will continue to control at least two segments not susceptible to LCCs. These are international travel, where service and network features retain value, and thin domestic markets, where diversified fleets can feed smaller hubs. Beyond that, it seems inevitable that a shakeout will occur among FSNCs, as their combined share of domestic markets declines.

Another intriguing trend among the major airlines in all parts of the world is the internationalization of air service through inter-corporate alliances. This is widely interpreted as the beginning of a global shakeout which will result in a small number of leading global carriers and a secondary tier of isolated national carriers. The chairman of Alitalia put the options starkly in 1987:

> smaller carriers have two choices ... (1) try to stay in the big league or (2) accept downgrading and try to retain a safe niche in a local market. The first option requires a carrier to embark on the road of associations, mergers and acquisitions, most likely reaching beyond the borders of its own country. (Gialloreto 1988, 199-200)

Alliances can be cemented through several mechanisms, distinguished by the degrees of integration they afford (Hanlon 1996). One type of limited association involves agreement among carriers to connect their separate route networks at shared terminal facilities. This may be enhanced by several measures including coordinating ticket sales (code sharing), schedules (joint numbering), and even seat purchase agreements (interlining) between separate carriers. The sharing of frequent flyer credits offers another variation. Wherever US-based carriers are involved, the question of anti-trust immunity has arisen (Feldman 1996). At the time of writing, three groups appeared to have achieved genuinely global connections. The Star Alliance brought together 13 carriers in 2001 including United, Lufthansa, and Air Canada, while the Oneworld group's eight members include American, British Airways, and CAI before its demise. The most recent of the mega-alliances is Skyteam, based on Delta and Air France (Solon 2002a; 2002b).

A tighter form of association, in legal and managerial terms, is based on cross-ownership through share purchases, or joint venture operations between carriers wishing to enter new fields. In North America, two cross-border equity alliances were struck during the crisis period of 1992, one linking American Airlines and CAI and the other involving Air Canada and Continental. Neither one flourished. However as a general type, these may include closer forms of operational integration such as sharing not only CRS systems and common maintenance pools but facilities at the respective hubs. By carefully forging key connections, the airlines can partly transcend the restrictions of bilateral agreements, skipping the conventional multinational stage. In effect "paired or cooperating groups of multinationals are now the state of the art in market penetration and dominant configurations" (Gialloreto 1988, 160).

Strategies

Air Canada: The Life and Times of a Flag Carrier

In Canada, airline operations are about many things, but perhaps the most critical factor separating success from failure has been access to secure, long-term lines of credit. Historically, the two leading air carriers were backstopped by larger institutions: railroad companies and governments in particular. When this ceased, in the 1980s, the dominant airlines found themselves in new territory, buffeted by the volatilities of financial and equity markets, business cycles, and extreme events. While the operating cultures of Air Canada and CAI (and their precursors) were never very similar, they did converge in certain respects over more recent decades. The collision of corporate traditions and changing circumstances reveals an interesting mix. For Air Canada, the central fact of its first 40 years was its privileged position as a state-owned instrument of air industry development.

THE CHOSEN INSTRUMENT

After a tentative start in the 1920s, Canadian commercial aviation collapsed in the face of the Great Depression. It was then that the Canadian government became concerned with the prospect of certain US airlines expanding northward and achieving dominant positions. Instead, Ottawa favoured the formation of a domestically owned firm to serve the home market for air transport. However, the choice of instrument proved politically complicated. Until the Conservative Bennett government cancelled air mail contracts in 1933, James Richardson's Western Canada Airways, based in Winnipeg, was the leading carrier. With the election of the King Liberals in 1935, plans were laid for the creation of a new Department of Transport, bringing together powerful agencies previously centred on rail and marine transport while carving out a new civil aviation authority free from the military. The designated minister, C.D. Howe, encouraged a new Toronto financial syndicate to compete for the mail contract as well. Liberal partisanship and bureaucratic manoeuvring, expressed particularly through the Post Office Department, worked against Richardson (Smith 1986). So did the British model of designating a crown enterprise, Imperial Airways, as the chosen instrument for aviation. Though Howe initially favoured a mixed state and private enterprise, the terms he outlined for its capitalization and governance proved unacceptable to private investors (Ashley 1963; Smith 1986). As a result, Trans-Canada Airlines (TCA) was established in 1937 as a state subsidiary of Canadian National Railways. While CNR head S.J. Hungerford was the founding president, the CEO was recruited from the Boeing-United conglomerate south of the border.

The new company operated within the terms of the Trans-Canada Airlines Act and according to a service contract negotiated with the federal cabinet. The air industry regulator, the Board of Transport Commissioners, which was replaced in 1943 by the Air Transport Board, was obliged to license TCA according to the terms of this contract, though the board also shaped TCA's prospects indirectly by regulatory decisions on applications by competitors such as Canadian Pacific Airlines (CPA). Trans-Canada's financial needs were met by an interest-bearing bond held by Canadian National. The government guarantee against loss was extended to 1942, after which Ottawa used the air mail contract as an instrument of adjustment for annual losses. These arrangements placed the minister and deputy minister of transport at the nexus of an elaborate policy network that included the department, the board, the crown corporation, and the other airlines. For the first 20 years, this worked generally to TCA's advantage for, as long as the Liberals held office, the airline found a powerful political ally in C.D. Howe. Shortly after the war, in fact, Howe had hoped to leave politics to assume the job of airline president, but he was persuaded to remain in cabinet until 1957. In a sense, this emergent air policy network was personified by the composition of the TCA board: four directors came from the parent CNR board, along with the CNR/TCA president, while three government directors came from the Post Office and the Department of Transport's Air Services and Civil Aviation branches.

When Howe failed to gain the presidency in 1947 it passed to G.R. McGregor, the airline's traffic department head who had successfully chaired the founding of the International Air Transport Association. The tone for the next two decades was set when McGregor visited Howe to clarify his terms of reference and was told "you keep out of the taxpayer's pocket and I'll keep out of your hair" (Smith 1986, 123). McGregor always insisted that Howe's formal interventions into managerial affairs were few and unsubstantial. The two men were in regular communication, and they shared an overarching ambition for the airline's success. Although TCA's growth record was impressive, with passenger volumes doubling every five years, it was also a period of sustained strategic rivalry. Under the vigorous leadership of Grant McConachie, CPA persistently lobbied the transport minister for international routes in the Pacific, Europe, and Latin America and approached the Air Transport Board for domestic rights to the lucrative transcontinental service.

For TCA, political difficulties began after the Diefenbaker Conservatives took power in 1957. In partisan terms, the airline's prior association with Liberal governments suddenly became a liability. In policy terms, the Howe dictum was seriously challenged by the new transport minister's commitment to competitive services. George Hees hired a British consultant, Stephen Wheatcroft, to explore the potential for domestic competition and also signed new bilateral agreements which brought new foreign carriers to Canada. Influenced by Wheatcroft's report, the Air Transport Board awarded CPA a foot in the door in 1958, with one daily flight and a maximum 12 per cent share of transcontinental traffic.

From this point on, policy confrontations multiplied, though not all were strategically based. The federal Department of Finance proved sticky in approving TCA's 1960 budget, which ballooned under the cost of new jet acquisitions. In the end, the prime minister's consent was required. Personally, Diefenbaker was antagonized by a series of minor yet symbolic steps such as the 1960 proposal to adopt the trade name "Air Canada" and the plan to shift a major maintenance facility from Winnipeg to Montreal. (Both were subsequently delayed.) An advertising contract with a Liberal-tied agency was terminated under pressure from the Conservatives. During an air policy review of 1961-62, a potential merger of the two carriers was seriously considered but dropped in the face of opposition from CPA. Following the Liberal Party's return to office in 1963, McGregor wrote indignantly to Transport Minister Pickersgill, denouncing "the infanticide that the Diefenbaker government was perpetrating on a government conceived and sponsored enterprise" (Smith 1986, 250).

The mid-1960s were a relatively prosperous time for TCA/Air Canada. It reported modest profits in the midst of a massive aircraft modernization campaign. However, McGregor's imminent retirement posed complicated questions of management succession, which had not arisen for 20 years. In Ottawa there seemed a new urgency to ensure the company's fit within national air policy, which implied a more politically sensitive leadership. This was reflected in both the proposal to split the jobs of chairman and president and in the decision to look beyond the McGregor lieutenants for successor personnel. Already Pickersgill was preparing to leave the cabinet for the job of chair at the new Canadian Transport Commission, and John Baldwin, the deputy-minister of transport, was angling for an Air Canada position. Both were confirmed late in 1968 when Quebec lawyer Yves Pratte was named chairman/CEO and Baldwin was named president. It marked the beginning of what one analyst labelled "the most calamitous upheaval in the airline's history" (Smith 1986, 266). To transform a management structure that had gone unchanged for more than two decades, Pratte hired McKinsey Consultants, whose report paved the way for a massive 1970 reorganization. In contrast to the McGregor philosophy of Air Canada as a regulated public utility dedicated to superior service standards, Pratte viewed it as bloated, inefficient, indifferent to profit, and in desperate need of business discipline. Not surprisingly, he sidelined an entire generation of senior management in forging his new team. The combined impact of organizational redesign, loss of executive continuity, declining morale, and macro-economic shocks — fuel and inflation — made the early 1970s a turbulent time, to say the least (Langford 1981).

COMMERCIALIZATION

This coincided with the first major political review of Air Canada's mandate since its founding. It began with a minor political furore over the airline's ties to a number of charter tour holiday operators. However, the ensuing judicial review raised more general questions about a lack of strategic direction and effective management and Air Canada's tendency to abuse its protected position (Canada 1975). In response, the Air Canada Act of 1977 authorized the recapitalization of the company, terminated its cabinet contract, and brought it under the same CTC regulatory control as its commercial competitors. Once the cabinet lifted its previous capacity restrictions in 1979, CPA applied almost immediately to the CTC for a cluster of new national routes, and the race was on.

Yves Pratte became a casualty of the inquiry findings when it became clear that he had lost the support of Minister of Transport Otto Lang. It fell to the new president, Claude Taylor, to steer the airline through the process of commercialization, which culminated in its privatization a decade later. In 1984, Air Canada's Annual Report forecast its capital needs for more than 40 replacement aircraft over the next decade and argued the need for new outside capital. After the Mulroney Conservatives took power with a firm platform to privatize crown corporations, the Air Canada board and management indicated its support for a public share issue. Despite this apparent synergy, the actual step toward privatization was delayed for several years. Initially it was complicated by the prime minister's public declaration that Air Canada was not for sale. Then the global stock market crash of October 1987 imposed further delays. It was not for another year that the enabling legislation was passed and the first block of shares (43 per cent) was issued, raising $246m, which went directly to the airline. Individual shareholdings were limited to 10 per cent, with an aggregate 25 per cent ceiling on foreign ownership. The balance of Air Canada shares were sold in October 1989, raising another $490m, which flowed to the federal government in return for its prior equity investment. By the end of the year, one of the largest federal crown corporations, with both assets and revenues exceeding $3b in 1987, had been transformed into a publicly traded business corporation under management control.

The newly privatized airline rested on a diversified foundation, in part the legacy of decades of favoured regulatory treatment. It held a steady 50-55 per cent of the domestic scheduled market, which included a dominant share of the leisure travel market and a dominant position in air cargo. This included a comprehensive network of regional "connector" airlines, with particular strength in Ontario and the West. At the same time it drew almost half of its revenues from international operations, where it held over 90 per cent of the Canadian-borne traffic on trans-border routes to the US. However, the first signs of a new cyclical downturn in air transport were also apparent by 1990, and as a high-cost, full-service carrier Air Canada was dangerously exposed.

Louis Gialloreto contends that, in the contemporary airline industry, there is "no space for the middle of the roader." He predicts that the latter, caught between hyper-niche carriers and multinational mega-airlines, would find that their ability to compete would decline relentlessly while their market shares would fall likewise (Gialloreto 1994). Air Canada qualified as a classic middle-road case. During the 1991 recession it pursued a conventional strategy of cutting costs and fares simultaneously. On the domestic front, it intensified the battle for market share, with chief rival CAI. The frequency and scope of seat sales and discounting more than matched their US counterparts. While domestic route revenues could never improve in such a context, Air Canada seemed willing to engage CAI in a war of attrition, gambling that its cash reserves would outlast its opponent's. The fallout was bitter, as both carriers recorded progressively larger losses in 1990, 1991, and 1992. In addition more than 7,000 airline industry workers were laid off (4,000 of them at Air Canada) during the 1990-92 period. Yet the autumn of 1991 saw Air Canada announce record seat sales on both domestic and international routes. The company also retrenched by eliminating non-core businesses. The central Montreal offices were sold, and headquarters staff were relocated to the airport suburb of Dorval. The *En Route* credit card business was also sold in 1992. While delays were forced by declining returns, the continuing program of fleet modernization brought fuel-efficient aircraft such as Airbus A320s and Boeing 757s into service, while older models like 727s and DC-9s were sold. This coincided with a retrenchment of marginal international services and emphasis on the core routes (Pooley 1996).

At the same time Air Canada was looking abroad for growth potential, especially to the US. In contrast to the caution of CAI, Air Canada was a strong supporter of the open-skies approach to renewing the Canada-US bilateral agreement (discussed in more detail below). This was seen as the most effective way to gain wider market access through one or more US hubs. However, delays in reaching a new bilateral agreement pushed Air Canada into a search for code-sharing allies, as a secondary strategy. In mid-1991 it reached tentative agreement with US Air, the fifth largest American carrier, only to have this come apart with the news of a larger British Airways-US Air deal. Like other airlines, Air Canada has found that the choice of optimal partners is a difficult business. After Hollis Harris arrived as president early in 1992, he took a 25 per cent equity stake in the sixth-ranked American carrier, Continental. He gambled that Continental could return from bankruptcy to financial health and anchor a North American alliance. Despite the code-sharing advantages this opened in the trans-border market, it was superceded in 1994 by a new code-sharing agreement with United Airlines, prompting Air Canada to divest its Continental stock, at a profit. In 1997, United and Air

Canada joined SAS, Lufthansa, and Thai Airways as founding partners in the global "Star Alliance."

Along the way, CAI's increasingly desperate financial position added another complication to Air Canada's strategy. (This issue receives more detailed treatment below.) It should be noted that the airline recession of the early 1990s brought unprecedented levels of bitter rivalry between the two carriers. The prospect of CAI bankruptcy, or of its rescue by American Airlines, forced Air Canada to respond. As CAI held talks with American Airlines through the winter of 1991-92, Air Canada mounted an intense lobby in Ottawa to block the deal. The positive spin in this campaign, presented as a "made in Canada" option, called for a merger of the two domestic airlines as an alternative to a *de facto* foreign takeover, which would turn CAI into a "junior partner" to a megacarrier (Campbell and Pal 1994). The negative current was a sustained campaign of predatory marketing aimed at crippling Canadian Airlines sooner rather than later.

It is clear that Air Canada was significantly transformed. Once a sheltered national flagship in the era of regulation, it was forced to become more commercially oriented and politically sophisticated. The key events in this story are highlighted in Table 4-2. Air Canada took the lead in assembling its regional connector network during the initial phase of *de facto* deregulation, thereby pre-empting independent entrants and fencing in its major competitor. In addition, it made a determined plunge into the emerging field of international alliances, with several abortive deals prior to the Star Alliance affiliation taking hold. Similarly it moved immediately to exploit the opportunities afforded by the 1995 open-skies agreement, seeking to build a set of profitable new routes before US carriers acquired unrestricted access to the big three Canadian hubs in 1998. All of these strategic steps pale, however, by comparison to the scramble set loose by CAI's impending bankruptcy in 1999. This extraordinary episode is explored in depth below. First, however, it is essential to consider the story of this formidable commercial rival, which pushed Air Canada so strongly during the 1980s and 1990s.

Canadian Airlines International and its Antecedents:
The Private-Sector Alternative

In the 1920s, when commercial aviation began in Canada, it was Winnipeg capitalist James Richardson who pioneered the leading firm. His Western Canada Airways was formed in 1926, using bush aircraft to move freight, carry mail, and offer rudimentary passenger service across the prairie and

Table 4-2

Key Events: Air Canada and Canadian Airlines

Trans Canada Airlines / Air Canada
1930
1937 TCA established as crown subsidiary — CNR S.J. Hungerford, President 1937
1940
Herbert Symington, President 1941
G.R. MacGregor, President 1947
1950
1960
1965 TCA renamed Air Canada Y. Pratte / J. Baldwin, CEO / President 1968
1970
1975 Air Canada Inquiry Report 1977 Air Canada commercialization act
1980
P. Jeanniot, President 1984 1986 Ottawa agrees to AC privatization 1988 First tranche of AC share issue
1990
H. Harris, President 1992 L. Durrett, President 1996 R. Milton, President 1999 1999 Onex bid and CAI takeover
2000
2000 AC begins to implement CAI merger 2000 New regulatory package (C-26) in Ottawa 2001 Terror attacks and air system closure 09 / 11

Table 4-2 (continued)
Key Events: Air Canada and Canadian Airlines

	WCA / CAL / CPA / CP Air	Pacific Western Airlines / CAI
1930	J. Richardson, President 1926 1930 WCA merger into CAL 1936 CAL fails in bid for transcontinental license	
1940	1942 Canadian Pacific buys CAL L. Unwin, President 1942 1944 Howe aviation policy denies RR ownership G. McConachie, President 1947	1947 Central BC Airways established R. Baker, President 1947
1950	1948 CPA wins international routes to Asia 1955 CPA wins first European routes 1958 CPA wins a daily transcontinental route	1953 Renamed Pacific Western Airlines 1950s Major DEW Line contractor in north 1958 Swaps transcontinental route for CPA's route network in NWT
1960	J. Gilmer, President 1965 1968 CPA wins 2nd transcontinental route	D. Watson, President 1958 1964 PWA begins vacation charters 1966 PWA designated BC/Alta regional carrier
1970	1974 Ottawa reallocates international routes I. Gray, President 1976 1979 Ottawa opens full mainline competition	1974 Alberta government takeover of PWA R. Eaton, President 1974 1977 PWA buys Transair
1980	D. Colussy, President 1982 CPAir buys EPA 1984, Nordair 1985 D. Carty, President 1985 1986 CP Ltd. begins sale of subsidiaries	1983 Albert government privatizes PWA PWA buys stakes in Time Air & Air Ontario 1986 PWA buys CPAir
1990	1987 PWA/CPAir renamed Canadian Airlines International 1989 CAI buys Wardair K. Jenkins, President 1991 1992 business crisis and AMR investment K. Benson, President 1996 1996 Second major employee/creditor concessions deal 1999 CAI predicts impending collapse, asks Ottawa to invoke s.47	

northern economy. A merger with eastern partners resulted in Canadian Airlines Ltd. (CAL) in 1930, with the CPR as a minority partner. Despite the impossible conditions of the Depression, CAL made the transition from bush aircraft to enclosed, all-metal planes.

For its first half century, air transport in Canada was closely regulated by government, and market prospects were defined, first and foremost, by federal policy. There were five crucial moments of policy definition: 1936-37, 1944-46, 1958, 1966, and 1977-83. The ability first to shape these decisions and secondly to respond to them is the story of private airline development. Key events are highlighted in Table 4-2. As mentioned earlier, Richardson fought vigorously for the transcontinental license that C.D. Howe announced in 1936. When it went instead to the state owned TCA, it was a bitter setback (Piggott 1998).

This defeat forced CAL to focus upon expanding its network of regional services. During the war, the company was instrumental in developing the Atlantic ferry route by which thousands of warplanes were flown to Britain. Then in 1944, following the death of James Richardson, CPR bought out the family interest and the firm was renamed Canadian Pacific Airlines (CPA). While this financial backing was crucial in such a risky and unproven business, no company in history has carried more political baggage than Canadian Pacific Railway. After decades of fierce battle in rail transport, Ottawa was determined to avoid the same experience in airlines. Consequently, Howe's aviation policy of 1945 blocked CPA from both transcontinental and international operations, the two most lucrative segments of the industry. Moreover, it required both railways to divest their airline subsidiaries within one year. This directive was ultimately reversed, however, when it became clear that TCA could not function without the umbilical tie to the CNR.

For CPA, this was a mixed legacy at best. The 1947 appointment of Grant McConachie as president opened a new strategic era. Until his death in 1965, McConachie was a forceful presence in guiding CPA (Keith 1972). Armed with a vision of growth, an appetite for risk, and the support of the parent railway company, he made the most of the limited set of opportunities that Ottawa allowed. McConachie's principal strategy was to go international. He persuaded the government to grant CPA several pioneering routes to Australia and Asia and, later, to secondary destinations in continental Europe, Latin America, and the US. Following the election of the Diefenbaker Conservatives, with their aversion to TCA, new political horizons seemed to open, and CPA won a single, daily, domestic transcontinental license.

Of course this was never a straightforward process, but depended on a combination of boldness and patience, along with substantial parent firm sponsorship (Goldenberg 1994). Over time, CPR presidents blew hot and cold on the air sector. In the 1940s, William Neal had been strongly supportive, while in the 1950s, Buck Crump was far more sceptical and withholding. In the 1970s, Ian Sinclair was again sympathetic.

The Liberal government's 1966 Regional Air Policy brought another mixed blessing for CPA. Classified as a national carrier, it was given a 25 per cent share of the mainline market. At the same time it was precluded from developing feeder markets by the separate licensing of the tier two regional carriers. Once CPA reached its ceiling, there was nowhere to grow until this restriction was lifted in 1979. In addition, Ottawa was not overly supportive of its applications in the international route review of 1974.

Appointed president in 1982, Dan Colussy made an important mark by reorienting CPA's strategic focus toward domestic growth. Since his tenure coincided with the *de facto* deregulation of the Canadian air market, it was Colussy's task to reposition his airline. It is important to remember both the pace and levels of change after Axworthy's 1984 proclamation. Windows of business opportunity opened and closed with striking speed. Entry regulation was first relaxed and then abolished. Air Canada and PWA were being readied for privatization, and competitive pressures south of the border were pushing fares through the floor. For CPA and Colussy, the room for manoeuvring was tight. CPA chose to structure its operations around hubs in Toronto and Vancouver, and this made more urgent a longstanding problem with feeder traffic to the mainlines. The initial answer was found in code-share deals with regional operators Eastern Provincial Airlines, Nordair, and AirBC. The ultimate response was CPA's takeovers of Eastern Provincial Airlines and Nordair-Quebecair.

In 1985, Don Carty assumed the CPA presidency and continued the strategy of growth through acquisition. The most immediate effect was the dramatic rise in debt, which reached $600m by 1986. In addition, the airline faced massive challenges of utilizing, and rationalizing, the patchwork fleet of aircraft that resulted from these deals. The airline faced urgent needs to renew the fleet with fuel-efficient aircraft. For example, the Boeing 747-400 was a full 25 per cent more fuel-efficient than the earlier model 747-200. In addition, CPA found that growth of market share was hard to achieve, with two other new mainline carriers (PWA and Wardair) also in the field.

Another uncertainty was the shifting strategic outlook of the controlling shareholder, CP Ltd. and its level of interest in airlines. This came into question in 1987, when the parent holding company began selling off subsidiaries in resources, hotels, and transport. While at one time CPA would have been regarded as a crown jewel, CP Ltd. had received no cash flow from its airline over the entire decade of the 1980s (Piggott 1998).

In the end, the problem was solved by divestment. In a striking reverse takeover, PWA acquired a carrier three times its size. Its 1983 privatization had left it with no debt and considerable cash on hand. A sale and leaseback deal on a block of planes provided an additional $250m, and PWA was, by 1986, responding to the deregulated market by going on the prowl for deals. Discussions were held with Air Canada, involving a merger-takeover possibility, but this would have to wait until after the Air Canada privatization was complete, in 1989 at the earliest. So attention turned to CPA; the $300m purchase was complete by the end of the year. But who was this upstart challenger and where did it come from?

THE PWA STORY

PWA originated in the bush-flight era after World War II, servicing isolated destinations in western and northern Canada. It began in 1946 as Central British Columbia Airways, one of the many small regional companies of the time, which later absorbed Queen Charlotte Airlines and Associated Airways. The name change to Pacific Western Airlines occurred in 1953. The onset of the Cold War brought a surge of defence-related business to northwest Canada, which PWA exploited from its base in Edmonton. The DC-3, formerly the workhorse of its fleet, gave way to the Bristol Freighter and the Hercules transport, with the DC-6 handling inter-city passenger traffic.

By the 1960s PWA's scheduled route map connected large and medium-sized cities from Calgary and Edmonton to Vancouver and Victoria. Then two key events paved the way for a major expansion in northern service. In 1966 the government of the Northwest Territories was moved from Ottawa to Yellowknife, bringing thousands of civil servants and a sizeable year-round passenger traffic. Two years later the Prudhoe Bay oil discovery marked the start of an arctic coast petroleum boom that continued for more than a decade. In both respects, PWA's hub at the Edmonton Industrial Airport offered a strong centre for the new surge in cargo and passenger trade. The new series of Boeing 737 jet aircraft were well adapted to this role and formed the core of the firm's fleet in the 1970s. PWA was thus one of the best positioned "regional" carriers in a decade of rapid growth.

The 1970s were also the decade of the "New West," a period of remarkable economic prosperity in Alberta and British Columbia. While this is normally associated with dynamic and high value resource industries, the

secondary and tertiary linkages, including an enlarged provincial and municipal public sector, were primarily urban-based. Indeed, the Calgary-Edmonton
Airbus became one of PWA's most profitable routes. Increasingly these factors tilted the company's business toward Alberta and the north. In a sudden
and radical stroke, Conservative Premier Peter Lougheed nationalized PWA
in 1974, moving its headquarters from Vancouver to Calgary. Beyond this,
the government was content with an arm's-length relationship that left senior
management in clear control (Tupper 1981).

During this time there emerged the executive team who would guide PWA
for the next 20 years. Rhys Eaton, a chartered accountant who had joined the
company in 1967, had worked primarily in northern operations. After the
Lougheed takeover he was appointed Executive Vice-President for Finance
and Planning. He was joined by Murray Sigler, a lawyer who had left a practice
in Yellowknife to join PWA. A decade later the operating environment had
changed markedly. The federal cabinet was carefully but persistently promoting the deregulation of the air sector. By the time PWA was privatized in 1983,
Eaton and Sigler held the number one and number two positions, respectively.
The Alberta government's strategy followed the British Columbia Resources
Investment Corporation format, by which equity was raised in a public share
issue that limited individual holdings to a modest few per cent of the outstanding stock. As a result, PWA became a widely held Canadian company, directed
by a management enjoying enviable cash reserves.

This presaged a decade of frenetic corporate expansion, which ultimately
put the entire firm at risk. PWA capitalized on Ottawa's new "open route"
provisions by starting trunk services from Calgary and Edmonton to Toronto.
The decisive move came in December 1986 with the takeover of the far larger
CPA. Over the next several years, the firm struggled to integrate three distinct companies (including Eastern Provincial Airways), realign its fleet by
substituting wide-bodied jets for the workhorse 737s, and remain price and
service-competitive in order to hold its vast new (35 per cent +) share of the
national market. At the same time, it continued to extend the regional air
"partner" arrangements already begun by CPA.

In the spring of 1987, a new operating company known as Canadian
Airlines International was unveiled as a subsidiary of PWA Corporation.
Further consolidation was pursued under conditions of intense competition,
since Wardair had shifted from charter to scheduled trunk service at the same
time and proved quite successful in bleeding CAI's traffic on the Montreal and
Toronto to Edmonton, Calgary, and Vancouver routes. As the market increasingly resembled its US counterpart, all firms reported losses by year end.

Perhaps the fateful move for PWA was its January 1989 takeover of Wardair. Occurring on the eve of an industry recession, it cost $250m in share purchase and added another $700m of long-term debt. Despite staff layoffs and airplane sales, PWA recorded a 1989 loss of $56m, its first in 19 years. The firm was caught in a classical recessionary dilemma: efforts to cut seat capacity risked loss of market share, sharp price competition prevented any revenue gains, and debt service payments mounted inexorably.

THE LONG DENOUEMENT

By 1991 PWA sought salary and work rule concessions from its unions, while it grounded planes and laid off 1,600 staff. In one apocalyptic interview, new PWA President Kevin Jenkins warned that, without change, the company could be dead within a year. By autumn two radical remedies were under consideration: partial or entire merger with Air Canada or alliance and equity stake by a foreign airline. For CAI, an era was ending. The great acquisitor of the Canadian airline sector had succumbed to its own ambitions. A way was open for survival, though in a severely rationalized form.

The industry had plunged into its first major recession since the advent of deregulation. The year-over-year trend in 1991 saw Air Canada's traffic decline by 17 per cent and CAI's by 8 per cent. Indeed, world air traffic declined in 1991 for the first time. This was exacerbated by the ballooning fuel prices and collapsed traffic associated with the Persian Gulf War. By the close of 1991 the very survival of the Canadian domestic air industry was in question. While the firms staggered from one loss-making quarter to the next, struggling to attract bargain-hunting passengers in a depressed market, two strategic solutions were advanced for the failing duopoly.

CAI was first off the mark in 1991, identifying an investment alliance with American Airlines as a lifeline to the future. Air Canada responded with its "Made in Canada" solution of a domestic merger resulting in a single carrier. For the next two years, events swung back and forth between the two like a yo-yo. Air Canada launched a major lobby campaign in the spring of 1992, only to see CAI break off to concentrate on the American option. The CAI-AA alliance seemed to have firmed up by the summer of 1992, only to falter over last-minute US conditions which Ottawa could not meet. The autumn was devoted to CAI-Air Canada merger negotiations, before several versions were rejected by the respective corporate boards. Early in 1993, cross-border talks resumed with American Airlines and received regulatory approval from the National Transportation Agency in May. In one of the last decisions of the

Mulroney era, the federal cabinet upheld this decision on appeal. However, a subordinate dispute stretched on for another nine months and postponed the consummation of the deal. It concerned the legality of CAI withdrawing from a long-term commitment with Air Canada in the jointly owned Gemini CRS system. After hearings in the spring of 1993, the Competition Tribunal concluded that it lacked jurisdiction to decide this question, triggering further litigation which only ceased when Air Canada dropped its action early in 1994 (Campbell and Pal 1994). Shortly thereafter, Ottawa announced that Air Canada had been licensed as the second carrier to serve Japan, in a move that some interpreted as a quid pro quo.

Perhaps the most extraordinary aspect of this policy debacle was Ottawa's inability to broker a solution in an obvious crisis situation. This contrasts strongly with its approach to the steel trade disputes and failing productivity in the pulp and paper sector. Campbell and Pal (1994) offer a number of incisive analytic comments on this. In the air case, Ottawa's hand was frozen in part by the difficulty in choosing between western and eastern champion firms, each with a bloc of popular regional support and each integral to the Conservatives' electoral coalition. This, however, is nothing new in Canadian national politics, and it did not prevent the Mulroney government from making other difficult choices. The broader legitimating paradigm of the Mulroney years played an equally important role, as did the ambiguities of the shifting air policy regime. Campbell and Pal point out that the neo-conservative commitment to market forces imposed a powerful brake on state sectoral intervention for embattled industries. Far more than in earlier eras, the political discipline of the government was being tested by this experiment in deregulated commerce. Somewhat triumphally, Ottawa had surrendered the traditional policy tools and found itself ill-equipped as a result. Repeatedly, the minister of transport could only appeal for an "industry-based solution" which the market proved unable to broker.

In this looming policy vacuum, the defective policy design of air deregulation was also starkly revealed. The resulting gaps and overlaps contributed to the confusion and opened a seemingly endless potential for litigation. This is the deregulated world of "quasi-politics" where conflicting interests battle for tactical advantage before quasi-judicial agencies in courtrooms (Campbell and Pal 1994, 86-87).

At CAI, Kevin Jenkins won labour concessions in 1991. Backed by investment from the AMR Corporation, the parent company of American Airlines, CAI managed to juggle debt payments, route structures, and aircraft acquisitions and was buoyed by the economic recovery of 1993. However,

Air Canada posed aggressive competition on all routes, and the emergence of low-cost discount carrier Westjet, in 1996, put extra pressure on CAI's western Canadian trunk lines. An extraordinary accounting write-down that same year led to Jenkins' resignation and replacement by Kevin Benson. A second major employee concession package followed, and the war of attrition continued. CAI was in no position to benefit from the Canada-US open-skies deal to the same extent as Air Canada. In retrospect, the mid-1990s was the last opportunity to rejuvenate the airline, but Ottawa was in no mood to raise the foreign ownership ceiling on which this probably depended. By the spring of 1999, the writing was on the wall, and Benson informed the minister of transport that bankruptcy loomed before year's end. The final chapter of the CAI story is explored in the second policy case, below.

Policy Issues

Canada-US "Open Skies": Free Trade in the Air?

We have already seen that the international airline industry has evolved since World War II within a policy regime quite different from that of domestic air services. As early as 1919, the Paris Convention confirmed that states controlled the sovereign air space above their territories. There remained, however, the considerable task of coordinating commercial movement across these air frontiers. In 1944, an international conference brought together 52 countries in Chicago to consider a multinational agreement on the freedoms of the air along with a regulatory system for fares, frequencies, and capacities on international routes. This did not prove achievable, given the unbridgeable distance between the "free trade" or "open skies" advocates, led by the US and including some smaller, outward-looking air nations such as the Netherlands and Sweden, and the more defensive European states, such as Britain, France, and Germany, who faced the virtual reconstruction of their air industries. This underlines the inevitable conflict between differing styles of passenger air policy. The international air sector varies tremendously in its importance to nation-states, being crucial to some but marginal to others. Furthermore, a nation's international route network may vary from the simple, involving only a single primary port of entry and exit (e.g., Holland or Hong Kong), to the complex, involving multiple ports and an equally diverse amount of "behind and beyond" traffic (e.g., the US or Canada). Such factors have a direct bearing on the definition of interests and negotiating strategies that national air industries and nation-states bring to international forums.

Figure 4-2

International Route Rights (Canada-US)

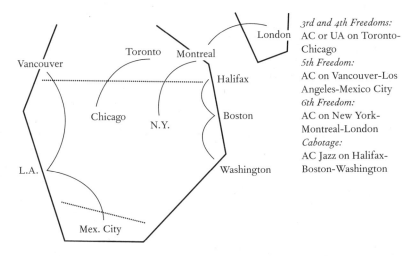

3rd and 4th Freedoms: AC or UA on Toronto-Chicago
5th Freedom: AC on Vancouver-Los Angeles-Mexico City
6th Freedom: AC on New York-Montreal-London
Cabotage: AC Jazz on Halifax-Boston-Washington

Freedoms of the Air

Negotiated in Bilateral Air Services Agreements (ASAs)

First Freedom: The right to overfly another country without landing.
Second Freedom: The right to make a landing for technical reasons (e.g., refuelling) in another country without picking up/ setting down revenue traffic.
Third Freedom: The right to carry revenue traffic from own country to treaty partner country.
Fourth Freedom: The right to carry traffic from treaty partner country back to own country.
Fifth Freedom: The right of own country aircraft to carry revenue traffic between treaty partner country and other countries (subject to agreement of other countries).

Supplementary Rights

Sixth Freedom: The use by own country aircraft of two sets of third and fourth freedom rights to carry traffic between those countries but using home base as transit point.
Cabotage Rights: The right of own country airline to carry revenue between two points in treaty partner country.

The Chicago meeting did approve the far more modest International Air Services Transit Agreement. It accepted the first two "freedoms of the air" (see Figure 4-2)—the right to overfly a signatory's air space as part of authorized service and the right to land in a signatory country for technical reasons, but without accepting or releasing passengers there. No agreement was reached on the exchange of the commercial air rights stipulated by the third and fourth freedoms; that is, the rights to carry revenue traffic from a home market to treaty partner market and back again. Neither was there any progress on the fifth freedom, which permits a carrier from one signatory market to fly traffic from a treaty partner market to a third market. The Chicago Conference did not even anticipate a further set of traffic rights that have become pressing in recent decades. These include sixth freedom rights linking two treaty partners on routes passing through the home market and seventh freedom or cabotage rights—flying passengers on route segments originating and ending in treaty partner territory. The latter two have yet to be accepted widely in the negotiation of agreements.

In this respect, the Chicago Conference foreshadowed the challenges faced by the comprehensive trade liberalization efforts which followed World War II. The ambitious US proposal for an International Trade Organization to promote multilateral tariff cuts on traded commodities proved unrealizable. Instead, a more modest, and more complicated, beginning was made at Geneva in 1947, with the General Agreement on Tariffs and Trade (GATT). Just as the GATT process began with a series of bilateral (two nation) tariff agreements between signatory states, the international air transport industry was built on bilateral air service agreements (ASAs). However, in contrast to the GATT, which evolved by the 1960s into a multilateral process, the air regime remains today bilateral in nature. Three key elements are involved:

> The exchange of traffic rights became a matter for bilateral agreement between states; the control of capacities and frequencies became a matter for inter-airline agreements, and sometimes for bilateral state agreements; and tariffs came to be regulated by the International Air Transport Association. (Doganis 1991, 27)

Over almost half a century this has created a patchwork of more than 1,650 agreements of varying terms and durations. The "Bermuda" Agreement negotiated between the US and United Kingdom in 1946 became a prototype for postwar ASAs. Some, such as the US-Japan bilateral agreement signed under the occupation in 1952, remained untouched for decades. Since it conferred generous landing rights on US carriers with few reciprocal benefits, Japan pressed persistently for its overhaul while Washington, not surprisingly, proved indifferent. Other bilateral agreements have been subject to periodic revision, as with the US-Canada agreement of 1949, which was expanded in 1966 and updated further in 1974 and 1984. As international agreements, they remain in force at the pleasure of the signatory states. Consequently, when France was unable in 1991 to reach agreement with the US over regulating an explosion of new US carrier traffic to Paris, it renounced the 1946 bilateral agreement and forced *ad hoc* seasonal negotiations in its place.

Each bilateral agreement is negotiated between national authorities, with varying degrees of involvement by the national airline industries. Closely affected companies may be consulted in advance, and both firms and trade associations may observe the negotiating sessions and advise their respective governments. However, the actual negotiations tend to be the preserve of foreign ministries, with technical input from ministries of transport, and the talks are typically conducted in the closed settings preferred by diplomats. In such contexts, state officials are free to weigh a spectrum of political interests, including consumers and regional economies as well as producers (Lissitzyn 1964). In addition, there is a strong possibility of linkage to non-air foreign policy issues, made greater when political leaderships become directly involved. In the 1957 renegotiation of the US-Germany bilateral, the State Department was preoccupied with its wider relations to a valued North Atlantic Treaty Organization ally, and the agreement was signed during Chancellor Adenauer's visit to Washington. Although US carriers were highly displeased with the resulting terms, they were relegated to the margins of this policy network. As will be seen below, the Canada-US open-skies agreement of 1995 was hammered out by two senior negotiators in a series of intense meetings and was signed less than two months later during President Clinton's visit to Ottawa.

Once an ASA defines a set of reciprocal traffic rights, it is up to the signatory states to allocate services among their respective air carriers. Canada, for example, chose in the early years to award all international routes to its flag carrier. This allowed TCA to claim route rights on any services authorized under bilateral agreements. Beginning in 1950, CPA was able to secure several overseas routes to Australia, Japan, and Latin America, but only after TCA had declined an interest. Then in 1964 the minister of transport assigned geographical spheres of operation, with CPA receiving the South Pacific, South America, Eastern Europe, and Amsterdam, while Air Canada was given the United Kingdom, Western Europe, the Caribbean, and most of the US. While designated carrier status could, in principle, be withdrawn and reassigned, this was not a common practice. As a result, a designated carrier can seek permission to suspend services or to pool a service with a treaty partner carrier (i.e., sharing the scheduled services or interlining—selling seats on one another's flights) in cases where low traffic levels discourage the full exercise of designated route rights. Both CPA and its successor CAI followed this practice on their slow-developing Asian and Latin American routes. Ottawa's tolerance for underutilized rights ended in 1994, when the minister of transport declared a "use or lose" policy and promised to open bids to competitors on such routes.

Traditionally, a bilateral ASA sets the parameters of the international airline business. It enumerates the routes to be flown (e.g., the Canada-US bilateral agreement lists more than 50 city pairs) and the fare structures which apply (normally by reference to International Air Transport Association schedules). The principles of reciprocity for designated carriers of both states and equitable benefit—balanced traffic shares—are normally acknowledged. Beyond this, the terms can be extremely detailed, as with the "Bermuda II" agreement between the US and the United Kingdom (1978), where efforts at capacity control actually stipulated the levels of traffic and designated which US carriers would hold rights to the strategic Heathrow airport in London. Conversely, in the more liberalized climate of the past decade, the US has promoted, with varying degrees of success, an open-skies ASA concept of unrestricted bi-national access to all routes, with full price competition.

CANADA-US OPEN SKIES

By the early 1980s there was considerable interest on both sides of the border to revise the Canada-US bilateral agreement. Over the years, this had grown into an impressive package, governing "the most lucrative, complex and comprehensive bilateral air relationship in the world" (Canada 1991, 6). In addition to the 1966 ASA, it included a 1974 Exchange of Notes that expanded scheduled service and specified point-to-point routes, as well as further agreements covering charter services and authorizing customs "pre-clearance" at certain airports on both sides of the border.

Much of the impetus for renegotiation stemmed from changes in the industry. The 1974 deal was built on the traditional logic of specified city pairs, which were viewed as "origin and destination" routes on which travellers from, say, Toronto wanted to fly to Chicago. Any continuing travel beyond this was seen as a matter for the domestic air network. Consequently, each city pair tended to be considered in isolation, as a discrete route. However, the international air business had in many ways outgrown such assumptions. Both the business and tourist segments were booming, larger jet aircraft were the norm, and carriers faced increasing pressures to fill those planes with passengers. This applied even before the deregulation of the US domestic industry. After 1978, the new hub and spoke system offered a growing structural advantage to the larger US carriers who could exploit their dominant metropolitan border hubs as well as feeding traffic to destinations behind and beyond.

Canadian competitors on the trans-border routes found themselves at a growing disadvantage, since their flights terminated at the US border city. Ongoing passengers were forced to "interline," or change to a US carrier on arrival, with all of the ticketing, baggage, and terminal complications this implied. This represented a considerable competitive handicap for the Canadian carriers. Furthermore, the existing trans-border route network was far from symmetrical in its penetration of the two national air markets. It gave US carriers direct access to 90 per cent of Canada's population, while Canadian carriers reached directly only 30 per cent of the US. The difficulty was further compounded by an emerging bias in the system of customs pre-clearance on Canada-to-US flights. This allowed Canadian passengers travelling on US carriers to arrive on the same basis as US citizens, avoiding customs on arrival and accelerating their onward connections. After examining the impact of these differences, the House of Commons Special Committee concluded that, for customers faced with a choice of airlines on a major trans-border route, "the incentive to use US carriers is overwhelming" (Canada 1991, 8). Recent traffic data appeared to confirm the pattern. By 1990, Canadian passengers constituted 62 per cent of trans-border passengers, but Canadian carriers attracted only 43 per cent of them, along with a similar proportion of trans-border revenues.

On top of these concerns stood the growing profitability of international service. By the late 1980s, the revenue return here was significantly higher than on the intensely fare-competitive domestic services in both countries. As one CPA executive put it, "There's potential to make money internationally; there's no profit domestically" (Potter 1986).

The first efforts to catch up culminated in 1984, when the Canada-US bilateral relationship was modified by the signing of two agreements. The Regional, Local and Commuter Service Agreement sought to inject greater competition in smaller trans-border markets served by smaller aircraft by enabling easier entry. The Experimental Trans-Border Air Services Agreement aimed to bring more business to two sample underutilized airports—Mirabel and San Jose—by allowing unrestricted, market-based service. While the thrust was toward liberalized access, both deals were aimed at peripheral rather than core services.

Of greater consequence was the round of bilateral talks launched in 1984-85, at the initiative of the US Department of Transport. Senior US officials, including Assistant Secretary Matthew Scocozza, were committed enthusiastically to the concept of "open skies." However, in the US industry opinion ranged from sceptical to hostile, coloured by the cut-throat fare wars, razor-thin profit margins, and chronic over-capacity that characterized the first phase of domestic deregulation. There were additional concerns that the commercial gains from open skies would accrue disproportionately to Canadian carriers and that a Canadian precedent would make it difficult to maintain restrictions on Japanese and European access.

The next opportunity to address the open-skies question came with the negotiations on the FTA (1986-88). The air industry was mentioned specifically in the "Shamrock Summit" communiqué that launched the free trade negotiating process. Although airline issues were discussed and addressed as part of the draft text, they were deleted at the eleventh hour, caught in cross-pressures from other issues. It was the US marine shipping industry, particularly the segment concerned about Canadian competitive advantage on the Great Lakes, that successfully forced the elimination of the Transportation Annex to the Agreement (Doern and Tomlin 1991, 86).

In the meantime, Ottawa pursued limited forms of air liberalization with other nations. In the autumn of 1987, Canada signed a new bilateral treaty with the United Kingdom. With relatively few stipulations, this allowed carriers from both countries to set their own fares and was described as innovative for its time.

By 1990, Canada and the US seemed prepared to try again. In Ottawa, Transport Minister Jean Corbeil appointed a Task Force on International Air Policy. Its five-volume report was submitted 14 months later, by which point the talks were already well advanced. At the same time, discussions between Ottawa and its two international carriers revealed significant differences in outlook. Air Canada supported the widest possible open-skies package including unlimited cabotage, which would allow it to build its own hubs in the US market. On the other hand, CAI opposed cabotage and argued the case for a more modest broadening of the range of approved city pairs. These positions reflected differing ambitions and abilities to immediately exploit the US market. With its stronger foothold—80 per cent of the existing trans-border routes—and stronger balance sheet, Air Canada sought to press its advantage, whereas CAI remained several years away from an international expansion campaign.

During the fall of 1990, a House of Commons Special Committee was asked to investigate the open-skies question and to consult with interested parties. After three months of extended deliberations, the committee endorsed the goal of liberalization, but warned that any acceptable deal would have to expand trans-border services while redressing the present revenue imbalance experienced by Canadian carriers. It also argued that phase-in arrangements and safeguard provisions would be critical to judging any bilateral deal. The committee declined to choose between the route-specific and the open-skies approaches, preferring to stress the need for acknowledging and protecting the special Canadian attributes in any policy outcome. Among the possible provisions that might help offset the Canadian carriers' structural disadvantages, the committee discussed a "limited" Canadian cabotage right allowing one intermediate stop en route to US destinations and the granting of fifth freedom rights to Canadian carriers on routes continuing to Latin America or elsewhere (Canada 1991). While it fell well short of articulating cabinet policy, the special committee played a valuable role in establishing the tilt of the playing field for international air transport and in triggering a public debate on Canada's air sectoral economic interests.

On the US side, the industry's reticence was being overcome partly by a ruthless market-driven shake-out of domestic competitors. A two-tiered system was emerging, with a strong elite resting on the Big Three — American, United, and Delta, joined perhaps by US Air — and a more precarious middle range of carriers struggling to reduce operating costs and avoid total failure. As it was, 1990 saw the disappearance of no less than three major carriers — Eastern, Pan Am, and Midway — with several others reorganizing under the shelter of the bankruptcy courts after filing for "Chapter 11" protection. Such differences of fortune and outlook created a greater opportunity for the open-skies initiative. In Washington, the Department of Transport was especially eager to conclude at least one strong open-skies bilateral as a model agreement that could be copied elsewhere. For this the Canadian market, evolving more slowly but in similar directions to the US market, must have offered a rather promising prospect.

The initial meetings were devoted to laying down groundwork and opening positions. The Canadians sought to establish the higher costs of operating north of the border in fuel prices, tax treatment of aircraft leases, and differences in market power. CAI contended that overall operating costs were 15 per cent higher in Canada and that reciprocity in route swaps required one-and-a-half to two route awards to Canadian carriers for every one award to a US carrier. But Ottawa's position was to press for a staged transition to fully open skies. First, route access would be opened, then ownership rules changed (both countries then restricted foreign voting stock to 25 per cent), and finally the cabotage provisions could be implemented. Ottawa offered immediate access to all Canadian airports except the three main hubs in Montreal, Toronto, and Vancouver, which would be phased in over several years.

By June 1991, the Americans had set out their requirements as well. In a major shift from previous offers of liberalized access to expanded range of city pairs, the new proposal covered "any type of air service between cities in Canada and the US." Washington also proposed open fare setting, unless both countries vetoed a proposed fare; the end of capacity restrictions; and the opening of the charter market, hitherto subject to another bilateral agreement creating a one-way market dominated by Air Canada, by removing charter class booking restrictions. One of the most complete statements of its open-skies position was issued by the US Department of Transportation in the spring of 1992. The central principles were open entry and unrestricted capacity and frequency on all routes (United States 1992).

As the year progressed, it became clear that industry conditions were becoming increasingly volatile. The Gulf War had driven fuel prices up and depressed international traffic levels. In the end, 1991 marked the first year since World War II that global air traffic declined in absolute terms. In addition, each failure of a US carrier caused a scramble among the survivors to capture the prime international routes and the landing slots at over-congested airports.

During the fall of 1991, the pace of the talks slowed, at Canada's request. There were signs that the Canadian position was under review, influenced no doubt by the worsening financial position of both CAI and Air Canada. The Minister's Task Force had also reported on, and advised against, Canada accepting unconditional open skies, which was judged to be severely damaging for Canadian carriers. In a straightforward fight, it was suggested, lower US operating costs would inevitably prevail. The new Canadian thinking seemed to lean in a different direction. Rather than a radical opening of route and fare possibilities, Ottawa seemed to see merit in enabling carriers to form strategic alliances with complementary foreign airlines. Therefore, it proposed that the foreign ownership ceiling in Canadian airlines be raised to 49 per cent. To further facilitate adjustment within Canada, it suggested that Ottawa abandon the "division of the world" policy begun in 1964, permitting international competition between Canadian carriers and the sale between them of international routes.

Irrespective of Ottawa's evolving position at the table, business and political events completely overwhelmed the policy planning process from the latter half of 1991 (Skene 1994). Air Canada announced in August its agreement to explore an alliance with US Air, while CAI's deteriorating finances drove it to renewed talks with American Airlines. Over the next 18 months, the increasingly desperate plight of CAI forced no less than three rounds of negotiations with American Airlines and two rounds of merger talks with Air Canada. By the end of 1992, CAI had a firm deal with American Airlines, which would see an investment of some $250m in return for an equity stake of 33 per cent overall and a voting stake of 25 per cent. The integration of the two carriers would extend to code-sharing, joint maintenance, coordinating schedules, and, most critically, the addition of CAI to American Airlines' "Sabre" computer reservation system. This brought an instant network of Canadian and new international destinations to a US carrier with little traditional involvement in such markets. In effect, it represented a faster and cheaper alternative for the two companies to extend their reach, in contrast to the open competitive scramble of open skies.

Through much of this time, Air Canada was aggressively pressing its merger/takeover proposal for CAI while keeping its eye on prospective US allies. After its talks with USAir collapsed, as the latter struck a more comprehensive alliance deal with British Airways, Air Canada settled on an alternative partner. The fifth largest US carrier, Continental Airlines, was struggling to emerge from a 1990 bankruptcy protection order. In 1992-93 Air Canada helped make this possible, by investing US$85m in cash, assuming US$140m of corporate debt, and acquiring 25 per cent of the equity in Continental. In addition to code-sharing, maintenance contracts, and share-earnings with Continental, Air Canada acquired an instant "behind and beyond" network in the US. In both cases, equity investment offered a means to extend international networks, irrespective of the outcome of the open-skies round. Significantly, Air Canada described its link to Continental as a preliminary step to capitalizing on a new open-skies bilateral.

Although reports persisted of a possible ASA deal during 1992, the trademark open-skies features such as cabotage and open entry had fallen by the wayside. Despite the resumption of talks in October, it was evident that with such fundamental corporate convulsions underway, the Canada-US bilateral negotiations had slowed dramatically. In addition, the intensified presidential campaign had becalmed the Bush administration on all volatile foreign trade issues, and no sudden results were likely as the electoral cycle entered its final phase. However, none of these factors took away from the reality that a new international air paradigm had solidifed in Washington. This proved to be surprisingly durable, spanning the partisan distance between Republican and Democratic administrations. Then, no sooner was the US election completed than the Mulroney government entered the final year of its mandate, with a similar drag on air policy initiatives.

The US-Canada bilateral talks emerged from their hiatus in 1995. Although Washington had achieved its model open-skies agreement some years earlier with the Netherlands, there appeared to be few other takers. While it was expected that the Canada-US negotiations would be revived once the Clinton administration was installed, international air policy was eclipsed by the ongoing domestic industry crisis in the US. July 1993 saw the preliminary report from Clinton's National Commission for a Strong and Competitive Airline Industry. In calling for a multilateral traffic rights agreement, it indicated impatience with the slow pace of bilateral liberalization and asserted the indispensability of open skies for a full US recovery. The implicit agenda of open skies was a scarcely concealed goal "to span the globe — with US airlines" (Shenton 1993, 5). Significantly, the Clinton administration endorsed the commission's policy package early in 1994. That, coupled with the slow revival of traffic and revenues in both Canada and the US, proved sufficient to revive the talks. Already new trans-border routes were emerging from between the lines of the 1974 and 1984 agreements. Indeed, Air Canada launched an aggressive plan to serve new border city pairs with a new fleet of smaller regional jets. Here Air Canada capitalized on a loop-hole in the existing bilateral agreements, which exempted aircraft of less than 100 seats — below the size of the old DC-9 — from normal bilateral rules, particularly in serving smaller cities.

In the autumn of 1994, both Canada and the US signalled an interest in reviving bilateral negotiations. In a first step, the transport minister and transportation secretary each named senior delegates to commence informal discussions. On the Canadian side, Transport Minister Doug Young named Geoffrey Elliott, a businessman seconded from Noranda Forest Inc., to conduct the talks free from the traditional bureaucratic delegation. At the same time, Ottawa also indicated its frustration with the old international air regime and an interest in wholesale liberalization, possibly through a new International Civil Aviation Organization treaty. Within months, the chief negotiators had initialled a seven-page framework agreement, and the background briefings spoke of a major breakthrough. Following two more months of detailed talks, the 60-page final agreement was signed by President Clinton and Prime Minister Chrétien in February 1995, and the era of open skies arrived.

By the terms of the deal, Canadian carriers acquired unlimited rights to operate point-to-point flights in the US without limits on capacity, frequency, or aircraft type. There was also provision to expand the landing rights available to Canadian carriers by one-third to one-half at the key hubs of New York-La Guardia and Chicago. US carriers acquired immediate access to all Canadian destinations except the big three, which would be opened gradually over three years to 1998. Notably absent from the agreement were cabotage rights beyond the initial points of entry. Overall, the result was to offer liberalized access to both sides, with an initial advantage to Canadian carriers. It was a compromise that sought to reinforce industry recovery while shielding carriers on both sides of the border from comprehensive route competition. For the Canadian carriers, Air Canada gained a strong growth market, and CAI gained a necessary extension to its American Airlines alliance.

Since the agreement took immediate effect, the carriers moved with speed to seize the opportunities. Thirty-four new services were put in place by June 1995, and another 40 were under development, most originating with the Canadian carriers. With newly delivered aircraft available, Air Canada was the most aggressive, deploying the smaller, 50-seat Canadair Regional Jet on starter routes and the larger Airbus for heavier traffic. CAI covered most of its new US flights under its marketing agreement with American Airlines. Overall, the first six months of the open-skies era saw cross-border traffic increase by 16 per cent, with new non-stop flights to 21 cities. In the short run, the stakes were greatest for Air Canada, which gathered 20 per cent of its 1994 pre-agreement revenues from the cross-border trade.

CAI and American Airlines sought to further cement their alliance in the autumn of 1995 by applying to the US Department of Justice for immunity from anti-trust actions. For allied airlines based in nations linked by an open-skies ASA, this has become a routine sequel to the finalization of a bilateral deal. In effect, it protects the US partner from charges of collusive behaviour on international routes. However, this particular application triggered a vigorous opposition from Air Canada in a campaign orchestrated by open-skies negotiator Geoffrey Elliott from his new position as the airline's senior vice-president for corporate affairs and government relations. The CAI-American Airlines forces argued, successfully, that their combined 14 per cent share of the cross-border market (5.4 per cent by CAI alone) posed no threat to Air Canada's dominant 25 per cent share, and immunity was granted in Washington early in 1996.

The trans-border market proved to be a lucrative segment for both Canadian carriers, but particularly for Air Canada, given its greater capacity to take up service on the new routes. Figure 4-3 indicates that trans-border traffic, measured in numbers of passengers, jumped 25 per cent in the period following the open-skies deal. While the terms of the Canada-US bilateral agreement have not changed since 1995, there has been sporadic pressure to revisit the questions of sixth freedom rights and cabotage.

Over the first eight years, the open-skies agreement achieved the expected effects. New non-stop services proliferated, adding 17 of the 50 largest trans-border city pairs, measured by passengers served in 2003. Ticket prices dropped an average of 22 per cent in these same 50 pairs. There were also

Figure 4-3

Air Passengers by Sector, Canada, 1986-2001

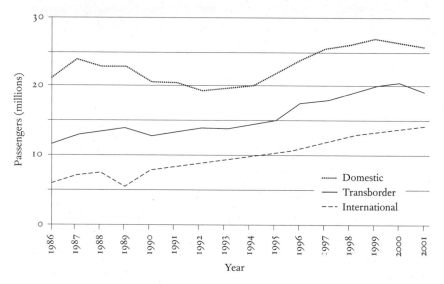

significant structural changes to the trans-border industry during this time. Four years into the deal, CAI ceased to operate, nullifying its alliance with American Airlines while its routes passed to Air Canada. Also during this time, Air Canada joined the Star Alliance and developed extensive code-sharing arrangements with Star partner United Airlines. More recently, the dot.com business collapse and the cyclical downturn beginning in the year 2000 meant reduced traffic on many routes—Nortel alone was reported to have spent $100m on trans-border travel before its 2001 decline.

In 2004, it is unclear whether the Canada-US air services agreement is complete or open to further extension. Washington has indicated a willingness to negotiate "full open skies" by which it means the inclusion of cargo service as well as the adoption of full fifth and sixth freedom rights to origins and destinations beyond the trans-border market. Air Canada has called for the inclusion of "modified sixth freedom rights," enabling a Canadian or US carrier to compete in east-west continental markets via a home market link, as in Boston-Toronto-Seattle (Dufresne 2003). Another option would see Canada focus on fifth freedom routes between Asia and the US, as in Hong Kong-Denver-Toronto (Elliott 2003).

The trans-border open-skies agenda may ultimately be reopened by events in Brussels rather than in Ottawa or Washington. In June 2003, the European Commission was given a mandate to negotiate a comprehensive trans-Atlantic

ASA with the Americans. Since the Europeans are pressing for the broadest possible open aviation area, issues of cabotage, procurement preferences, and foreign ownership rules are all likely to be raised. Consequently Canada, and Canadian carriers, may need to respond either pro-actively in a North American context or reactively to an emerging transatlantic package.

Onex and the Air Canada Monopoly: The Takeover/Merger Issue of 1999

As shown above, the Canadian airline sector has experienced a series of dramatic structural shifts in the period since 1980. The most recent and the most profound of these began in 1999, when the loosely regulated duopoly collapsed into a near monopoly. This was a complex and highly unpredictable process that pitted several corporate and state interests in a bitter and sustained struggle. Almost a year elapsed from the onset of this crisis to its definitive resolution. Even then, a continuing cluster of policy complications surround Air Canada's hegemonic position.

The hard-fought crisis of 1999 sets the context for today's industry. In exploring the business-government dynamic of contemporary air transport, we begin with the CAI failure and the policy responses it elicited. Almost at a stroke, the premise of a workable air duopoly was abandoned. Instead, virtually all new market-based solutions were premised on the emergence of a dominant, near-monopoly provider. The fall of 1999 saw an extended tactical skirmish among proponents of rival deals, followed by the political approval of the Air Canada takeover option. The next year and more was devoted, in Ottawa, to fashioning a policy framework commensurate with this new industry structure. This required major new legislative instruments and a renegotiation of state mandates among the federal transport regulator (the CTA), the sector bureaucracy (the Ministry of Transport), and the emerging lead agents (the Competition Bureau and Tribunal). These will shape the political agenda for air transport now and in the future.

THE TRIGGER

Cash shortages, surplus capacity, and book losses were nothing new to CAI as it entered the 1999 fiscal year. The Calgary-based carrier had lurched from crisis to crisis with every seasonal downturn, interest rate hike, or contract negotiation. The strong late-1990s upturn in Canada's air travel market (captured in Figure 4-3) may have prolonged the day of reckoning without altering the final outcome. Nonetheless, it became clear to Kevin Benson, in the spring of 1999, that his beleaguered firm was running out of funds. Based on prevailing trends, CAI's cash holdings were forecast to be exhausted before year's end.

This precipitated the search for yet another white knight, an investor willing to recapitalize the firm in exchange for a controlling ownership position. Initial discussions with the Onex Corporation, a Toronto-based holding company and leveraged buy-out specialist, began in this way, though no firm arrangements emerged. By the summer, CAI's position had moved from urgent to desperate. In July, Benson met with federal Transport Minister David Collenette and revealed the extent of the damage. The transport minister made it clear that his government was not prepared to orchestrate a financial bailout and that a "private sector solution" would be required. The industry minister, John Manley, called for an "orderly restructuring" that would stand the test of time (Fitzpatrick and Jack 1999). A special cabinet committee on airline restructuring, including the ministers of transport, industry, and justice and with Privy Council Office support, was struck to develop a federal response.

In early August, CAI asked Ottawa to open the way for it to conduct wide-ranging merger and takeover discussions without risk of prosecution under the Competition Act. This, it was suggested, required the use of s.47 (the "extraordinary disruption" clause) of the Canadian Transportation Act, which authorized the cabinet to take any steps necessary to stabilize the national transportation system. On Friday, 13 August 1999, following the close of the stock markets, Collenette announced a 90-day waiver of the Competition Act provisions that restricted the sharing of financial information between rival firms. This opened a window until mid-November for CAI to continue full service while it openly sought to reorganize.

THE CONTEST

In fact, CAI's impending collapse had been an open secret in the airline sector since at least March 1999 when Kevin Benson began quiet talks with Toronto financier Gerry Schwartz of the Onex Corporation. Out of this came a bold new "one airline" strategy in which Onex, CAI, and its minority shareholder AMR would capitalize a takeover/merger with Air Canada to establish a single national carrier. This strategy was outlined in June, again in confidence, to Transport Minister Collenette and justified as the only viable private-sector solution. It remained, however, under close wraps for the next several months. Nonetheless, Onex began purchasing Air Canada shares through two numbered companies.

For its part, Air Canada shopped a very different plan to the transport minister in June 1999. This centred on a proposal to buy CAI's valued international routes to Asia and elsewhere, while allowing CAI's domestic network to "interline," or connect on favourable terms, with Air Canada's expanding global service. In effect, Air Canada sought to buy the best and conquer the rest. Once again, there was no inkling of these bold plans in the business press. Air Canada CEO Lamar Durrett announced his resignation on 6 August and was succeeded by President Robert Milton. In retrospect, it seems clear that Air Canada was designating a new team to wage the prolonged takeover and regulatory war that lay ahead.

Almost immediately following the 13 August announcement, a series of speculative solutions found their way into public debate. One version involved route realignments, with either Air Canada or CAI holding the strong hand. By contrast to Air Canada's proposal sketched above, the CAI version saw it sell domestic regional and even central Canadian routes while refocusing on international service, perhaps with routes added from Air Canada. A second version saw American Airlines increase its CAI equity stake above the 25 per cent ceiling stipulated by law, contingent upon a liberalization of the foreign ownership in transport rule. A third scenario involved a CAI bankruptcy to be followed by the establishment of new and debt-free carriers to bid for its 40 per cent market share. Fourth was a proposal to secure an unconditional open-skies agreement with the US, enabling CAI to share in this lucrative and expanding market segment. Rounding out the early speculation, the 1991 "maple leaf" scheme, involving a full scale Air Canada-CAI merger, was dusted off.

In less than a week, however, all options were eclipsed by the fanfare accompanying the public disclosure of the Onex strategy on 24 August. This took the form of a double purchase and merger plan, whereby Schwartz would, with the financial support of AMR, establish a holding company known as AirCo to buy control of both airlines and merge them into a single carrier. This deal, with an estimated value of C$5.7b, would combine $3.9b in debt finance and $1b in new Onex cash, with the balance provided by AMR. As described, the AirCo equity would be held by Onex with 31 per cent, AMR with 15 per cent, and private shareholders with 54 per cent.

The Onex offer to Air Canada shareholders ($8.20) and Canadian shareholders ($2.00) remained in effect until 8 November 1999. During this two-and-a-half month campaign period, a bewildering range of political interests bid for influence. The pre-scripted character of the CAI-Onex alliance was evident in the almost immediate acceptance of the proposal and positive recommendation to shareholders by the CAI board of directors. Not surprisingly, the Air Canada board dismissed the hostile Onex offer as of little merit. Indeed, it followed up by approving a poison pill defence against takeover and by scheduling a shareholders' meeting on the offer for 7 January 2000, well after the Onex bid expired. Nonetheless, the Onex "big deal" succeeded in defining the business and public policy context for the next three months and rendered all earlier proposals marginal by comparison.

Over the following months, each of the key actors manoeuvred to advance its core interests. The Onex share tenders targeted both boards of directors and shareholders, particularly the institutions that held an estimated 80 per cent of Air Canada stock. The courts also emerged as a tactical theatre, as various sides asked judges to counter the gambits of opponents. For example, Onex sought an Ontario Supreme Court action to require the Air Canada shareholders meeting to take place before the 8 November expiry date on its tender. Air Canada replied by asking the Federal Court of Canada to declare that any merger deal must be subject to review by the federal Competition Bureau. Most significantly, in the end, was Air Canada's second action in the Quebec Superior Court, contending that the Onex offer was illegal in violating the 10 per cent maximum shareholding ceiling in the Air Canada Public Participation Act.

Though the federal government had hoped to stand aside for the 90 days of business matchmaking, this was not possible in practice. In early September, Ottawa came under fire for ignoring the public interest and allowing a policy vacuum to emerge. For example, the role of the Competition Bureau was unclear. Industry Minister John Manley had asked not for a review of the Onex proposal but only for an advisory opinion on air competition issues in general. Furthermore, despite the plethora of transport policy issues posed by Onex—ranging from foreign ownership limits to Air Canada shareholding limits to foreign carrier competition to the protection of consumers and small markets—Ottawa's silence was deafening. After weeks of continuing pressure, Transport Minister Collenette set out five conditions that would guide Ottawa's final decision on, and relations with, a dominant carrier. They included protecting consumers against price gouging, ensuring continued service to small communities, respect for the rights of employees, ensuring competition within the industry, and the continuance of effective domestic control of air services. Following the recall of Parliament in October, this was amplified with Transport Canada's release of *A Policy Framework for Airline Restructuring in Canada* (Canada 1999b).

Much of the policy confusion derived from uncertainty about the meaning of the s. 47 action in August. Was it merely a legal adjustment to allow wide-ranging talks without fear of prosecution? Or was it intended to create a fast-track process in which the cabinet would vet the ultimate proposal on general political grounds of public interest acceptability, free of the narrower technical criteria of competition, equity ownership, or service standards? It seems clear that CAI and Onex saw the s.47 action as a facilitating measure. Equally, Air Canada saw it as an obstacle that had to be overcome by legal action. As the weeks and months unfolded, whatever unanimity the cabinet enjoyed in summer gave way to doubts and nervousness in autumn. As well, the opposition parties raised charges of cronyism between Collenette and Schwartz, since the minister had arranged for the businessman to take a flight in a Canadian military F-18 several years earlier, and Schwartz's Liberal affinities were a matter of record. Even some Liberal backbenchers showed discomfort with the bland face of federal power.

Air Canada's counterattack was dedicated to blocking the success of Onex's share tender in the market while weakening the political case for AirCo as a secure and stable solution. Air Canada mounted an aggressive buy-back for 35 per cent of its rapidly appreciating shares, in an effort to deny Onex the level of control it sought. To meet the estimated billion dollar cost, Air Canada drew on its Star Alliance partners United Airlines and Lufthansa. Onex responded with its own enhanced share offers that were financed in part by its partner, American Airlines. The final Onex offer of $17.50 per share was more than double its initial $8.25 tender. Meanwhile, Air Canada was developing a bold alternative bid of its own, announced on 19 October. The mirror image of the "one big airline" model, only in Air Canada's favour, this involved the purchase of CAI and its operation as a separate subsidiary. It also proposed the merger of the rival regional carriers and the creation of a third unit, a discount airline modelled on Southwest Airlines. The latter concept had been studied by Air Canada and rejected earlier in the year, only to be dusted off as part of the grand counter-bid.

THE TIPPING POINT

In most decision-making processes, there comes a decisive moment when one possible outcome gains ascendancy over its rivals. Sometimes the crucial factor can be a modest or even marginal element, not obvious at the time, that serves to shift the balance of advantage between several complex and finely balanced alternatives. While there is always a danger in posing this question after the fact, since events tend to appear far more clearly and orderly in retrospect, it is still a useful and revealing question to pose of any policy process.

In the air merger case, it was in the Quebec Superior Court chamber of Judge André Wery that the outcome was decided, just days before the Onex tender was due to expire. As already seen, several strategic litigations had been launched that autumn on both sides of the battle. However Air Canada's challenge to the legality of the Onex tender, on the grounds that its 35 per cent ownership bid for Air Canada equity violated the statutory 10 per cent shareholding limit, was potentially the most significant given its "veto" ramifications. If upheld, it would negate the Onex bid at its very foundation. The judicial question was whether it was legal for Onex to advance the bid, or legal under prevailing law for Onex to consummate it. By finding that Onex "proposed a course of action that was illegal," Judge Wery closed the door in a ruling that was widely described, in the business press, as "surprising" (Gibbens 1999). Even if leave was granted to appeal this trial verdict, it could not be resolved within the stipulated time frame.

Did the failure to account for this risk represent a lapse of due diligence on the part of Onex? Not necessarily. The future of the 10 per cent rule, in place since 1988, would inevitably figure in any comprehensive review of air transport policy, as would the 25 per cent foreign ownership limit in any domestic carrier. A discerning review of these and other air policy issues was provided by the Competition Commissioner in his September report to the minister of transport (Canada 1999a). Collenette's own framework paper on airline restructuring, released on 26 October, acknowledged as much. However, amid the throes of the takeover battle was no time to advance legislative amendments to the underlying policy regime. This would have been politically unsustainable in any event, given the extraordinary levels of hostility and recrimination prevailing between managements, shareholders, unions, politicians, and the public. Moreover, once the House of Commons Standing Committee on Transport began its autumn investigation into airline restructuring, any new policy initiatives had to await its report (Canada 1999c). Such factors underline the importance to the eventual outcome of the constantly shifting dynamics between 24 August and 21 December. So many configurations changed in this time. The cost of a takeover, and the debt implications for the victor, grew exponentially and alarmingly. The scale of Air Canada's preferred strategy began with a modest route purchase and rationalization scheme in August, but grew into a gargantuan three-carriers-within-a-company scheme by October. Even the federal government's reluctance to intervene during these months assumed a shifting significance. Initially, it implied a bias in favour of the Onex "private-sector" deal, but ultimately it metamorphosed into a bias against that deal.

THE POLITICAL LEGITIMATION OF THE "NEW" AIR CANADA

Resolution came at the close of the year, after six weeks of manoeuvring. With its rival eliminated, Air Canada attempted to trim some of the commitments made in the heat of battle. Ottawa pushed back, announcing its approval of the merger on 21 December 1999, conditional on a series of "undertakings" agreed by the airline. These included no staff layoffs for 24 months (to March 2002), the creation of several new Air Canada operating units (discount, charter, and regional) with varying degrees of autonomy to service separate market segments, the delay of the low-cost carrier start-up until September 2001 (should a rival discounter appear before Air Canada could launch), and the right of competitors to first refusal on any Air Canada/CAI planes put up for sale. Recognizing the sweeping implications of the merger and the political complications it set loose, Ottawa promised a full domestic air policy review by 2002.

Without question, the 1999 events redefined Canada's air policy template. For the federal government, the first crisis demanded a response to CAI's impending and terminal collapse. The second was to achieve a viable result among the limited business options available. The third was to cope with the emergence of a lightly regulated near-monopoly, while making it politically sustainable. While the first two appear to have been weathered, the latter is likely to remain a central political problem, today and in the future.

The Chrétien government advanced a series of low-threshold measures beginning in February 2000. Bill C-26 defined a more active role for the director of the Competition Bureau, invited more assertive fare reviews by the CTA with penalties where warranted, and offered the public cosmetic assurance through the establishment of several new watchdog "representatives." Former hockey referee and travel agent Bruce Hood was named Air Travel Complaints Commissioner, responsible for reporting retail customer feedback. Second, Debra Ward was appointed as the Independent Transition Observer for the two-year term to follow official approval of the merger. Finally, Transport Minister Collenette trumpeted moral support for new entrants into all air sectors. On the latter point, the record was decidedly mixed. The Halifax-based Canjet operated for a mere few months before selling out to Canada 3000. The longer-established Royal Airlines followed the same path. Toronto-based Roots Air, designed as a full-frills, executive-class service, lasted only a few weeks before being sold to Skyways Holidays. It seemed that several of the new entrants were little more than franchise fee instruments for their sponsors.

With the exception of Westjet, no major carrier made money consistently in air transport in the 1990s, and Collenette aimed to give the newly consolidated Air Canada a substantial period of grace. The minister signalled, in actions if not in words, that he would not re-regulate domestically, nor deregulate across borders, until a reasonable settling-in time had passed. Politically, the government sought to sail between the deregulated success stories of Westjet and Canada 3000, and the lurking Air Canada colossus with all of its intractable contradictions. Collenette also agreed to slow or block expanded access for foreign carriers. Canada would discuss reciprocal expanded sixth freedom rights with the US and the possible lifting of the 25 per cent foreign ownership rule, but only in time.

After he left the portfolio, Collenette said that air policy occupied more of his time than any other matter. If so, the minister certainly mastered the art of political self-restraint, demonstrating a determined willingness to be patient for a transition period of two to three years to allow Air Canada time to cope with its extraordinary organizational and financial challenges. While the consolidated Air Canada stood seventh among North American carriers and thirteenth in the world in 2001, it remains far from the stature of American or United Airlines.

Yet Robert Milton demonstrated huge ambitions for the re-engineering of his company. In the post-merger period, the strategic shape of the "new" Air Canada became clear. It was a bold corporate response to a situation that was both economically and politically unsustainable. Organizationally, the "two airlines in one" arrangement was never more than a halfway house. While routes and schedules could be rationalized, and aircraft sold or retired, huge personnel problems attended the efforts to merge the employee groups, a situation that was perpetuated by the two-year, no-layoff rule. Moreover, since both Air Canada and CAI were full service network carriers depending upon executive travellers for their profit margins, the plunging business demand in the 2001 market slump, exacerbated by the terrorist attacks of 09/11, had a devastating effect on revenues. In the process, Air Canada became a lightning rod for consumer wrath, rivalling the chartered banks as a prime target of popular hostility.

Milton's response was to redesign the formerly homogeneous Air Canada into a series of specialized companies in order to segment the commercial passenger market. Operating at varying degrees of arm's-length from the main airline, they provided organizational flexibility rather than rigidity and decentralized management that isolated the company elite from potentially controversial choices, such as abandoning marginal small city pairs or launching controversial "fighting brands" to undermine discount rivals. Furthermore, the newly established subsidiaries could escape the inflexible wage, seniority, and job classification terms that bedevilled the mainline parent. Finally the "family of airlines" approach paved the way for subsidiaries to be sold or launched as freestanding firms. The disaggregation of services aimed, in effect, to turn a hitherto undifferentiated industry into a series of subindustries. In Louis Gialloreto's terms, Air Canada sought to become a layered complex of types one, two, and three carriers.

Commercially and politically, the long-term prospects for Air Canada's multi-divisional or multi-company operating strategy remained uncertain in 2002. The consolidated regional carrier, renamed Jazz, reported weak financial results on a declining traffic base. The long-haul discount carrier Tango, whose entry was delayed due to Canjet's appearance until November 2001, made early inroads into the low-fare market, which continued to expand even as executive travel declined. A third niche business, Zip, commenced operation in the fall of 2002. Designed to challenge Westjet on its short-haul discount routes in western Canada, its operating cost structure is a full 25 per cent lower than the mainline Air Canada, as was Tango's. Debra Ward aptly characterized Air Canada's new approach as

> the "boutiquing" of airline products. From a full-service, high-cost airline, Air Canada is attempting to develop products that will meet the needs of all travellers: low-fare point-to-point (Tango), high-end (Jetz), low-cost networked (Zip) and full service and regional network (AC mainline and Jazz). Only time will tell if the strategic shift from a bum in every seat (yield management) to the new paradigm, a seat for every bum (niche products) will work, but trying to meet new demand with only high-end products clearly won't. (Canada 2002c, 6-7)

Whatever its virtues as a business strategy, this new model raised a host of policy complications for Ottawa. Just how long a period of grace could the minister allow? Given the slump in trans-border traffic since 2000, how far could this period of grace contribute to sector recovery? In what fashion would federal competition measures fill the policy vacuum, given that the line between aggressive rivalry and abuse of dominant position can be a fine one? In addition, Debra Ward's concluding recommendation for full continental liberalization, including foreign ownership and reciprocal cabotage, offered a durable new point of policy advocacy (Canada 2002c). This was even more compelling given that Air Canada enjoyed a market dominance not seen for a generation.

Conclusion

Appearances to the contrary, there is nothing simple about the scheduled passenger air transport industry. On close examination, it actually consists of two or three separate businesses—for domestic, trans-border US, and international services. While the same infrastructure and equipment applies to them all, the state policy regimes are sufficiently different to ensure that, from the enterprise perspective, each segment must be managed in its own way. This is also an industry that has experienced radical change in its structural coordinates over a single commercial generation. The regulated regime that still flourished as recently as 20 years ago is now little more than a distant memory. In the meantime, the scheduled passenger industry has seen an intense period of business consolidation, achieved through mergers and acquisitions; a decade-long phase of duopoly between privately owned rivals under loose state surveillance; a parallel phase of commercialization and privatization of air infrastructure, ranging from airports to navigation services; and finally an era of single-firm dominance and increased product differentiation within the corporate umbrella. No one who has lived through these developments, or who has studied them, could conclude that the air industry lacks drama and challenge. Analytically, it is a story of shifting policy networks and outcomes that carries many lessons for political issue management.

When the air industry is mentioned in Canada, it is the experience of the domestic segment that normally comes to mind. In its origins, the commercial air sector was heavily state-directed, with the central government utilizing its strong jurisdictional base to articulate a national interest in a Canadian-controlled air industry. This was accomplished, as we have seen, through efforts to promote a single domestic champion—private, mixed, or state-owned—in a regulated environment. While the outcome was a crown corporation with a decidedly privileged business status, there was still room for politically driven adjustment according to the party of government in Ottawa. The regional air policy of the 1960s and 1970s represented a further adjustment on the regulated model in the effort to broaden the range of operators around regional hubs. This, together with a cautious loosening of pricing rules and the blurring of the scheduled/charter divide, coincided with the boom in air travel associated with jet carriers and the leisure clientele.

Ultimately, a combination of consumer advocacy, new economic ideas, the US experience, and political entrepreneurship in Ottawa, led to the domestic market liberalization of the 1980s. This, too, displays a unique Canadian style and outcome. There was no bold stroke but rather a prolonged four- to five-year policy transition in which both firms and state authorities recalibrated incrementally. Contrary to expectations inspired south of the border, there was no flood of new entrants to challenge the established carriers, but rather a corporate takeover binge that led by 1989, to a national duopoly. This left the survivors encrusted in debt, administering high-cost operations, and locked into endless discount pricing battles. It led, not surprisingly, to the crises of the 1990s.

In retrospect, one of the most surprising features of domestic air deregulation was its policy naivety. Apart from the consumer welfare aspect, which clearly suggested an electoral advantage to the government that could claim credit, the politics of this process have the hallmarks of opportunism, incrementalism, and wishful expectation. Once the regulated industry paradigm was fractured by Transport Minister Axworthy in 1984, the way ahead seemed unclear. Having failed in the political battle of the early 1980s, the commercial air carriers lost both their network access and their coherence of vision, with every firm forced to fend for itself. Similarly, the air policy section of the transport ministry saw its preferred blueprint for limited fare flexibility rejected, and the transport regulator had its mandate reversed by ministerial directive. By 1985, the air policy network was in disarray.

A telling aspect of this was the gross failure to manage the interface between regulatory and competition policy. The blunt fact was that the dismantling of the regulatory regime created a vacuum that state policy needed to address. In the US the answer lay in a vigorous anti-combines regime that was extended to the newly deregulated airlines. To be fair, the more sophisticated champions of deregulation, such as the Economic Council of Canada, insisted that an augmented competition policy be part of the policy transition. In practice, however, Canada's historically flaccid combines regime was insufficient. It was, in fact, in the process of fundamental revision during the critical years of air deregulation (1981-86). With the competition policy paradigm in flux, and the competition bureaucracy suspended between old and new, it is not surprising that the flurry of critical airline takeovers went virtually unchallenged. Even under the best of conditions, the task of managing the shift between such divergent policy cultures would be daunting. Under the conditions of the 1980s, it proved altogether dysfunctional.

As we have seen above, the international air segment has followed a separate trajectory in recent decades. In some respects it is anomalous, both to air transport and global trade trends. The system of bilateral air service agreements has continued since the early postwar period, as states have been unwilling to surrender their mercantile licensing prerogatives. For their part,

firms have adapted by designing intercorporate alliances, which are, effectively, global networks that can transcend the ASA regime. This is not to say that
the substance of bilateral agreements is not changing. A prime case in point is
the Canada-US open-skies model, which points in potentially new directions.

Today, the air industry remains in a fluid state, to say the least. In some
respects, the structure of the domestic scheduled market recalls the conditions
of 40 years ago, when TCA also held over 65 per cent of the business share,
an overwhelming amount. This time, however, there is no state enterprise
instrument. It may well be questioned whether the federal government has
lost its capacity for effective intervention. Certainly the events of the past
decade show that Ottawa has lost the ability to set the air industry agenda.
The handling of the Onex-Air Canada battle suggests that the policy goal is
to maintain a functioning commercial service under domestic ownership. The
gradual emergence of significant competitors would be welcomed but is not
essential — perhaps analogous to the regional carrier role of the 1970s. In this
sense, the collapse of Canada 3000 was particularly unfortunate, as it seemed
ready to join Westjet as a rival to the Air Canada colossus.

How then might single-firm dominance be stabilized commercially and
legitimized politically? In business terms, Air Canada has revealed an innovative, though still untested, strategy of intra-firm competition delivered
through a series of differentiated subsidiary operations. Its aim is to delineate
a range of market segments that offers separate services under varying cost
and price structures. This offers several possible gains. It can be used to
counter the consumer or populist critique based on the dangers of a virtual
monopoly. It also offers concrete evidence of the differential performance of
business units based on full service and discount, long-haul and short-haul,
and union and non-union platforms. Finally, it offers an organizational
vehicle for spinning off select subsidiaries into free-standing firms, either by
tender sale or by public offering. Air Canada's regional carrier Jazz seems, at
time of writing, to be a possible candidate for this. Tango, however, has been
terminated as a carrier and now exists only as a fare class.

On the political front, the continuing difficulty of justifying single-firm
dominance is considerable. As part of the Air Canada approval of 1999,
Ottawa established the dual offices of complaints commissioner and transition
observer. Both, however, were time-limited responses to consumer and public
antagonism and can be fairly characterized as cosmetic in their impacts. In the
long run, the only defensible policy regime would seem to involve the vigorous enforcement of competition. Under the legislative amendments of 2000,
the federal Competition Bureau has made a start, and the Westjet-Canjet

complaint of 2001 offers an interesting test case. The settling-in of competition oversight will, however, extend to a far wider variety of issues. These may well include matters of non-price competition bound up in frequent-flyer programs, CRSs, and airline alliances.

The same may be said for infrastructure access and service costs, ranging from terminal gates to surtaxes and fees. The privatization of airport facilities in the 1990s has altered both the scope and structure of the Canadian air sector. Airport enterprises in large cities are now assertive players with distinct interests in the policy network, and they are insisting on formal recognition at the clientele level. They now actively promote increased traffic, expanded services, and new carrier operations at their hub facilities. Their collective trade voice, the Canadian Airports Council, has weighed in on a variety of policy issues. Early in 2004, for example, it mounted a campaign for Ottawa to sell federally owned airfield lands to the airport authorities, as an alternative to the present rental regime. The new air industry security regime is another instance of a policy problem that accentuates the both the interdependencies and the tensions between carriers and ground services.

The only other major policy option involves a significant redefinition of foreign carrier roles to inject rivalry into the Canadian domestic market. While Ottawa has steadfastly dismissed it as a viable option, it is telling that both the federal Competition Commissioner and the independent observer now publicly favour some degree of liberalized rights to foreign carriers. For the time being, however, the shallow base of air policy coordinates mean that both the merger legacy and the structure of passenger transport remain in doubt.

Pressing as these policy issues are, they were eclipsed in 2003 by the Air Canada bid for bankruptcy protection. This year-long interlude is a fascinating political encounter in its own right. Given its "stand-alone" nature, the story of the airline's most recent near-death experience is addressed in the Postscript below.

Key Terms and Relationships

Alliances: Inter-firm agreements to link networks to achieve global reach.

Aeronautics Reference: A key court decision of 1932 confirming dominion jurisdiction over air transport.

Bankruptcy Protection: A legal process in which courts temporarily protect a failing firm from its creditors and supervise a reorganization. Known in the US as "Chapter 11."

Bilateral ASA: A state-to-state air services agreement defining the permissable terms of reciprocal service.

Code-sharing: Agreement between carriers to list flights jointly and coordinate schedules.

Cross-subsidy: A central technique in regulated markets of using buoyant revenue streams from high-traffic routes to offset losses incurred on low-traffic routes.

Fit, Willing, and Able: The new standards of liberalized licensing under the National Transportation Act of 1987, which replaced "public convenience and necessity."

Frequent-Flyer Programs: Customer loyalty rewards, pioneered in the 1980s, to gain a non-price advantage in attracting and retaining passengers.

Freedoms of the Air: Rights of overflight and gathering and landing passengers as codified in the Chicago Convention and its amendments.

Hub and Spoke: The route structure emerging after deregulation in the US, in which an airline feeds its traffic from city pairs (spokes) to central nodes (hubs) and onwards.

International Air Transport Association: An international trade association for airlines that concentrates on global as distinct from national air policy issues.

Licensing authorities: Key state agencies in a regulated air policy regime, with control of routes, fares, ownership, and airworthiness (eg: in Canada: Air Transport Board, Canadian Transport Commission, National Transport Agency, Canadian Transportation Agency).

Maple Leaf Proposal: A proposal advanced by Air Canada to create "one big airline" through merger with CAI as a solution to the latter's impending failure.

Open Skies: Proposals to liberalize, in varying degrees, carrier entry and expansion in the domestic or international sector by rescinding regulatory controls.

Revenue Passenger Mile/Kilometre: A key business measure of return on operations.

Use or Lose: The principle that carriers who fail to operate licensed routes risk having them cancelled and reallocated to competitors.

Yield Management: The use of fares and booking terms to ensure maximum capacity use.

CHAPTER FIVE

Conclusion

Our goal in this book has been to explore Canadian business politics at the microlevel. We began by posing certain questions about how and where the interests of markets and states coalesced. Despite the dramatic transformations of both economic and governmental structures over the past half-century, these questions have lost none of their relevance in contemporary life. Business-government relations remain a central context for modern management and policy-making. The search for corporate political advantage takes place alongside, but distinct from, the pursuit of commercial advantage.

The politics of business power cannot be simply inferred from market structures or commercial chains. Neither, however, can such politics be understood in isolation from these economic foundations. Accordingly, a series of concepts and models were introduced in the first chapter in an effort to capture the multiple dimensions of our subject. Each of these sits at the boundary of the twin domains of market and state and highlights a connective relation: internal corporate politics, product cycle, industry structure, rent chain, policy network, and state capacity, to name a few. These are tools that can be applied, in principle, to any inquiry into business politics. Working with the cases of paper, steel, and airlines, we explored the impacts of historical precedent, production structure, corporate strategy, and policy conflict.

While each industry stands on its own for purposes of analysis, this short conclusion casts a comparative eye across the findings and offers some final thematic comments. The industrial classification includes literally hundreds of industry groups, only three of which have been tackled here. I hope that readers will be encouraged to launch their own investigations into business politics. The appendix offers some guidelines to available research materials, with particular attention to the worldwide web. But first, here are some closing reflections, organized around a set of summary propositions.

Industries are Politically, as well as Economically, Constituted

It is clear that industries have national and sometimes regional shapes, and these shapes are of the greatest political significance. Despite technologies that easily cross global boundaries and broadly comparable product outputs that move from one country to another, these business structures are not all cut from the same cloth. Our three Canadian examples serve as compelling warnings, even in an age of world business, against unwarranted universal inference. One key source of differentiation stems from the modes of governing industries. States exhibit contrasting policy styles in such areas as regulatory intervention, reliance upon state enterprise, procurement preferences, taxation practices, the management of trade, and approaches to competition policy. Stable and resilient production regimes, such as those identified for the Canadian paper, steel, and air transport sectors, are normally rooted in foundational policies imposed or agreed-upon by capital and state interests. The turbulence that arises in the absence of such policies is also a political product.

In Canada, the air transport sector has evolved in a manner that has few equivalents abroad, the closest perhaps being that of Australia. At first blush, it may seem that the air industry followed a familiar, almost classic, pattern. It began with a half-century of growth within a state-regulated framework, beginning in the Depression years. This then dissolved into a progressively liberalized regime, which was imposed by federal state authorities after 1984. Yet this gloss conceals many critical distinguishing features. The Canadian regulatory style was anything but classic, as a favoured, state-owned, flag carrier received special treatment by its contract with the federal cabinet. Furthermore, the initial industry regulator was a railroad agency that brought a pre-established mindset to airlines. For 40 years, successive regulators sought to shape the industry at its margins while lacking jurisdiction over the dominant transcontinental carrier. When the regime changed in the 1980s, the deregulatory process was not sharp and clean but tentative and incremental. For commercial carriers, this brought rivalry not from new entrants but a concentrated shakeout through merger and takeover. Ultimately, a cut-throat duopoly was resolved into a single-firm dominance, leaving the industry structure in 2002 looking strikingly similar to that of 1950. In Canada, these are the *differentia specifica* of airline politics.

In primary steel, a similar portrait of nationally framed politics can be drawn. Historically, iron and steel emerged in a tariff-regulated environment that dictated both the shapes and the flows of products. Following World War II, a mutually advantageous policy consensus was agreed-upon between

Ottawa and the steelmakers. Based on domestic specialization and scale, the adroit use of imports, and a relaxed competition posture, this resulted in a world-competitive steel sector by 1980. The relative strength of the integrated firms, measured in productive efficiency and profitability, made primary steel a prominent Canadian business success story. Steel was also the site of path-breaking trade union drives during and after World War II, at the same time as the imperatives of the planned economy laid the basis for future steel expansion. Except where the crown owned reserves of coal and ore, provincial governments lacked direct policy leverage in the steel sector. One way to remedy this was through equity ownership, an approach that Saskatchewan, Nova Scotia, Quebec, and Ontario have followed at different times.

So comfortably embedded was this policy-strategy consensus that Canadian steelmakers lacked even rudimentary associational structures for collective action when this era began to erode. The rise of global oversupplies of steel shifted the political focus in the 1980s to the political management of trade flows. This problem has proven intractable over more than 20 years, and, as one of the few regional economic blocs that remains a net importer of steel, North America has had a particularly difficult time. Almost all of the proposed responses are politically driven: resort to conventional anti-dumping and countervail remedies, the creation of integrated cross-border markets, expanded hemispheric trade zones, and multilateral agreements on capacity reduction and fair trade rules.

Finally, the paper industry yields a story of specifically Canadian business politics. The paper sector emerged in the late nineteenth century in response to the commercial demand for newsprint in the modern print media and for packaging in consumer goods. This involved complex chemical-physical and engineering processes for turning logs into pulp and pulp into paper. The boreal softwood forests were well positioned to supply pulp fibre for the US market, provided that crown timber owners offered favourable stumpage and tax ratios. While US tariffs worked to slow the export of finished papers but not of pulp, the powerfully organized US publishing association secured the early removal of the newsprint tariff, and this segment rapidly emerged as the export engine of the Canadian industry.

The broad distribution of forest and mill operations across Canada made pulp and paper manufacturing, unlike steel, a coast-to-coast "national" industry. Yet the regional contrasts between the eastern sector with its first-generation investment, older mills, and pulpwood input, and the western sector, which was characterized by second-generation investment, newer mills, and sawmill waste inputs, meant that the political geography of pulp and paper would be complicated, particularly where provincial authorities got involved. This industry displays strong patterns of political representation at both federal and provincial levels. The national trade association was one of the earliest to appear and became one of the largest business interest groups in the country. Its historic role was to vouchsafe the market as the prime allocative arena while managing government policy concerns at the margin. As offshore producers achieved competitive advantages in market pulp and newsprint production, firms were challenged to reposition themselves in higher value product segments. However, there is strong evidence that traditional Canadian strength in the commodity products, along with high levels of sunk investment, slowed the pace of this adjustment. By the 1970s, pulp and paper was one of the first industries to draw attention from the emerging environmental movement. Since that time, the burden of social regulation has been a continuing preoccupation, embodied in rules for solid and chemical effluent elements, bleaching agents, end-of-pipe treatments, closed loop plant processes, and recycled paper quotients.

Multiple Fields of Engagement

Industry politics operates at a variety of levels. Mihaescu's compelling image, of the man in the bowler hat, offers a discerning glimpse into the complicated weave of diversity and unity among business interests. An effective understanding of our subject must acknowledge these multiple domains and be sensitive to the interplay between them. One aspect of this multiplicity is institutional. Organized political pressure can be exerted through different agencies of state: elite politicians, legislative organs, bureaucratic politics, federal-provincial executive channels, and judicial dispute settlement, to name some of the most prominent avenues.

Any industry operates within a wider economy that can impinge in a plethora of ways. Similarly an industry represents a set of rivalrous producers which share, nonetheless, a set of common interests. For certain analytic purposes, however, the industry is too gross and unwieldy a unit. Here the role of the subindustry or "business," as Magaziner and Reich (1982) put it, also must be grasped. Finally, we have seen that each firm is a political as well as a commercial organization. The choices inscribed in corporate strategy and the policy perspectives of the company are a product, in part, of the firm's internal relations of authority. These multiple political force fields can be seen in our three industry cases.

TCA was given a privileged mandate and enjoyed significant though not unlimited autonomy in pursuing it. Special relationships with Liberal governments facilitated this arrangement, while Conservative interregnums complicated and ultimately terminated it. The lesser, privately owned, commercial airlines—Western Canadian, CPA, PWA, and the other regionals—were left to scrap for shares of the residual markets. Over time an alternate tradition emerged, one of growth and repositioning through corporate mergers and takeovers. It is hardly surprising that this trend continued, and intensified, during the era of liberalization. Often these consolidations were linked to provincial state interests for resident carriers, enhanced service, or critical infrastructure.

It is significant that Canada's domestic air regulation regime was transformed, in the 1980s, against the will of the regulatees. The parliamentary consultations made clear that neither individual airlines nor their collective voice, the CATA, were remotely in favour of deregulation. Rather the pro-market campaign was driven by a coalition that combined consumer interests, economic ideologues of the neo-liberal persuasion, and political entrepreneurs like transport ministers Axworthy and Mazankowski. Domestic deregulation may also have been sped by Ottawa's growing unease with air transport becoming a site of federal-provincial rivalry.

Another setting which posed inter-firm rivalry in sharp relief was international air service. For the airlines, foreign route assignment was a largely zero-sum contest with the federal government parcelling out the rights in a time-honoured mercantile fashion. This policy process depended on international diplomacy and foreign policy considerations quite distinct from the template for domestic air transport. Flag carriers could sometimes parlay the symbols of sovereignty into commercial advantage. However, the absence of air companies from the bargaining table, together with the inherent tendency for issue-linkage and diplomatic log-rolling, meant that international air remained a state-directed policy arena. A few "open-skies" models to the contrary, it is striking that the modern trend toward domestic market liberalization have not extended widely to international air transport.

The Canadian steel sector was built upon state bounties and tariff protection. Its development path was conditioned through the twentieth century by a further range of policy calibrations. The design of freight rates played a role, as did the tacit restraint of anti-combines review, tax incentives for capital investment, and the regulation of the labour movement and employer-worker bargaining relations. Each constituted a separate policy arena where Canadian steel interests were forced to contend with rival forces, ranging from foreign producers to domestic consumers to allied industries in the chain of commerce. Provincial governments perhaps played more of an assertive role in steel than in airlines. This began with their role as crown resource owners and extended to trade union law, equity stakes in peripheral steel plants, joint federal-provincial schemes for mill modernization, and procurement leverage in product markets like steel rails.

With the rise of the excess global supply conditions in the 1980s and 1990s, the political focus shifted to the regulation of the traded steel sector, with particular attention to import flows. At a stroke, agencies like Canada Customs and Revenue and the CITT became focal points for anti-dumping and countervail tactics aimed at defending domestic market share and strengthening pricing power. Similarly, Canadian steel manufacturers became intimately familiar with the political folkways of Washington's ITC and the Congressional Steel Caucus.

In pulp and paper, the political arenas can be distinguished by their roles in the origin, growth, and mature phases of the industry. Initially, it was secure access to forest fibre, on concessionary terms, that put provincial departments of lands and forests in the centre of the picture. However, prospective markets were also crucial in dictating the range of products and technologies employed. Here federal trade and competition polices became operative. Prior to the rise of the environmental movement, the political treatment of pulp and paper was production-oriented and focussed on the employment and export-earning contributions of this leading manufacturing sector. By the close of the 1960s, however, the policy frame began to change. The inherent competitive strength of Canadian mills could no longer be taken for granted, and concerns mounted about the viability of the older generation of predominantly eastern facilities. Even more significantly, pulp and paper was identified as a leading source of industrial pollution at a time when environmental degradation drew urgent public attention. An industry that was traditionally approached as an undifferentiated whole was increasingly illustrating internal fault lines based on age of plant, location, products, ownership, technique, and efficiency. Public policies for the older generation turned increasingly toward strategies for rationalization and closure.

Globalization's Many Shapes

Globalization is now recognized as an important but problematic reality. Without question, the world economy is today bound together in many ways—cross-border trade, capital flows, money transactions, communications, and cultural diffusion—to name only the most obvious. The transnational corporation is the exemplar of such activities, conducting cross-border business within a single enterprise. Today their role is unprecedented. Collectively, the world's transnational firms control no less than one-third of all private-sector productive assets. Similarly, state organizations such as the IMF, the WTO, the G-7 (a group of seven highly industrialized countries) and G-10 (that group expanded for reasons of international finance), and the OECD maintain rule-based structures to stabilize international business and standardize business policies. If nothing else, this has accelerated the speed with which international economic transactions occur and shortened the response times for firms, industries, and governments having to deal with the consequences.

Beyond this, however, questions have been properly raised about how much of these relationships is new. The proportions of international trade between nations today and the scale of cross-border capital flows are not significantly different from that of 90 years ago. Furthermore, the most authentic form of transnational production, in which chains of production extend across borders, is far less common than the headquartered parent with subsidiaries carrying on separate operations in multiple national markets. There is a difference, then, between "international" and "global" business. On the other hand, the scale and impact of the world money market represents a genuine globalizing force whose flows are uniquely capable of destabilizing the prospects for firms, industries, and states. In short, it is entirely appropriate to consider the ways in which global forces may be altering the political foundations of business in Canada. However, it is equally important to recognize that, even in an age of global reach, national boundaries and residency still delineate the fields of micropolitics.

The contemporary phenomenon of globalization can infiltrate industries in a variety of ways, and our three cases illustrate a surprising number of these. All three are quite heavily traded. In pulp and paper, this applies particularly to market pulp and newsprint and less so to tissue, packaging, and fine papers. For the traded segments, the international marketplace is a constant conditioning force. Shifts in market pulp supply and demand make this a highly cyclical industry and shade over into the pricing of other paper segments. Significantly, the pricing mechanisms include both contract deliveries and spot sales, with the latter registered on the new commodity and derivatives markets like Finland's FOEX and Britain's Livedex. In newsprint, the posted prices set by leading producers like Abitibi-Consolidated, International Paper, or Stora Enso are extremely sensitive to international supply and demand conditions. For some plants, a fluctuation of $25 to $50 per tonne can spell the difference between profit and loss. Further yet, money markets inject constant volatilities into currency exchange rates, and these can confer differential advantage or disadvantage on Canadian exporters or importers.

Significantly, the paper industry has been able to exploit the organizational advantages of transnational production far more effectively than the other sectors. This enables leading Scandinavian and North American paper firms to internalize key business transactions and utilize corporate strategy on a far higher plane.

In the steel industry we have a case where, prior to 1960, national markets were largely separate with very little traded account. Then the relentless growth of new capacity together with growing imbalances of domestic supply over demand gave rise to chronic global surpluses. In Canada and the US, one of the few leading regional blocs not self-sufficient in steel, this made the trade question the overwhelming strategic concern of the past generation. Here global pressures are largely transmitted through trade policy practices. This can take the form of market opening tactics such as aggressive exporting aided by concessionary finance, dumping, or state ownership, all potentially subject to countervail remedies. Given the vastly different cost-price structures that apply to northern Asian, western European, post-communist, and southern hemisphere producers, the range of fair and unfair trade permutations seems infinitely complex.

The international air transport sector illustrates the challenges of achieving growth and profit in the face of neo-mercantile bilateral arrangements. This, coupled with the continuing concern for reciprocal grants of flag carrier privileges, raises questions of whether an authentic traded air service "market" even exists at present. Nonetheless, national carriers have explored a number of innovative adaptations over the past several decades in their search for global reach. These include equity swaps (still widely restricted by statute), coordinated scheduling, and code sharing. Most recently this is seen in the formal business alliances that have been struck between groups of independent airlines, such as the Star Alliance, Oneworld, and SkyTeam. In effect, this pattern of internationalization at the level of partnered enterprise is an adaptive response to the continuing rigidities of the Chicago system and the continuing failure of the state system to agree on a more liberal multilateral regime. Should such alliances evolve into more formal governing structures, this may provide a platform for further business integration.

Policy Networks as Crucial Connectors

In situations where either the advocates or the targets of public policy assume a group dimension, associations can be expected to play a significant role. And since associations tend to connect with state authorities in persistent and structured ways, it is important to be sensitive to the impact of network relations. Our three industries display a number of network types that are surprisingly dynamic over time.

The scope and scale of the pulp and paper industry was set relatively early in the twentieth century, and a strong representational umbrella was fashioned in the CPPA. Its professional sections, annual meetings, journals and newsletters, and well-staffed Montreal headquarters offered a clearing house for a variety of industry concerns from woodlands management to paper engineering. On the production and commercial side of the industry, CPPA expertise won it ready access to federal authorities in key departments such as finance and industry, trade and commerce. This relationship was classically clientelistic and mutualistic. There were few rival associational interests of consequence. Business-government relations were characterized by a high and continuing level of technical exchange. So long as the pulp and paper sector remained commercially competitive, Ottawa viewed it as a privileged manufacturing sector. This effect can be seen in the industrial strategy exercise of the 1970s, when this sector proved one of the easiest to mobilize for consultation. For eastern Canadian producers, at least, the pulp and paper modernization program was a notable outcome. At the provincial level a different sort of clientelism emerged, not around matters of tax, trade, and innovation, but around primary wood supply and pricing. Concessionary leaseholds were an integral part of pulp industry development throughout the twentieth century.

The workings of these clientele pluralist networks were tested, however, by the emergence of the modern politics of pollution and the spectre of effluent regulations. This betrays a different pattern, one of greater state capacity in rule-making with a far wider range of non-business interest involvement. We have noted, of course, the sequence of effluent control regimes that followed the fisheries regulations of 1970. The first wave, with its generous grandfathering provisions, retained traces of the earlier clientelism. Also the industry tactic of blunting federal regulatory action in favour of provincial jurisdiction reflected a continuing expectation of producer-first policy. This was dislodged as the emissions issue was redefined by awareness of a wider spectrum of toxins and their possible health effects. The advent of Greenpeace, the Sierra Club, and other ENGOs brought a far more pressure pluralist cast to pollution politics in the 1990s. Most recently, there may be signs of a neo-clientelism associated with self-regulatory corporate approaches to environmental management.

In network terms, the steel industry offers a curious contrast. A smaller range of integrated firms, with far greater concentration at the upper end, this sector lacked the elaborate organizational infrastructure evident in pulp and paper. This does not mean steel was politically disinterested, unconnected, or unsophisticated. Far from it. This was an industry that had prospered under the wartime control administration and gained a significant reconstruction boost from the privatization of wartime plants and the new capital investment tax provisions. The Big Four also played an influential role in the Tariff Board's rationalization of duties in 1957. In short, the steel oligopoly enjoyed ready access to federal state circles on the issues that mattered, and it was able to do so without an elaborate associational vehicle.

In fact, the prime trade lobby for the leading integrated Canadian steelmakers was the American Iron and Steel Association based in Washington, DC. This symbiotic relationship between big Canadian and US steel continued through the postwar decades of growth and prosperity. It ruptured, however, with the North American steel trade disputes that broke out in the 1980s. Then the CSPA emerged, interestingly enough, to articulate the shared positions of integrated and mini-mill operators. Yet for reasons of timing as well as issue affinities, the network form appears to be quite different. The modern arena of import-export management may be inherently unsuited to clientelistic politics. Domestic steel producers find themselves arraigned not only against the mercantile interests affiliated to offshore suppliers but also against the domestic steel-consuming industries whose political as well as market power may be formidable. The Parliamentary Steel Caucus may number 35 members but the Liberal Party's Auto Caucus includes more than double that number. In the crowded and contradictory setting of trade policy disputes, all players face the challenge of manoeuvring in a pressure pluralist network.

As a subset of this, hemispheric trade relations continue to bedevil Canadian producers. During the formative negotiations of the 1980s, the steel sector ranked among the most serious supporters of a comprehensive Canada-US deal that would clarify trade rules and dampen the threat of continuing friction. Yet from steel's point of view, the incomplete character of the resulting agreement left almost all anxieties outstanding. Today the sector's unease about the shortcomings on subsidies and dispute settlement can be seen in its reaction to CITT rulings on dumping, countervail, and safeguards. Should the trade policy network for steel remain vibrantly pluralistic, as seems likely, firms way well turn to market-based strategies for relief. For some this will involve IPSCO-style diversion into the US; for others it may involve cross-border mergers or alliances along the lines of Co-Steel/Raritan.

In air transport, the central political fact before 1985 was its legally regulated character. State capacities were strong, and carriers were closely constrained by the combination of cabinet contract for Air Canada and CTC supervision for the remainder of the industry. There may be room for debate on whether the clientele or concertation network best captures this reality. The carrier interests were clearly the dominant constituency, interacting with a regulator that displayed marked "capture" characteristics. But the stability of the transactional framework and the ongoing support for its purpose and outputs suggest more fundamental joint planning. Either way, it resulted in a striking political consensus between regulators and regulated firms, a convergence that continued up to the events of 1983-87.

Then, in several sharp strokes, the underlying alliance dissolved, and the domestic air policy network was fractured. The goals of stability and profitability were replaced with flexibility and competitiveness. The sectoral principles of "public convenience and necessity" gave way to "fit, willing, and able." Most critically, a political and policy vacuum developed in the wake of domestic deregulation. Ottawa's competition policy regime, which should have offered a more supple and sensitive framework, was itself in the process of fundamental revision and was not an available option. Meanwhile, the minister of transport reflected the psychology of disengagement that gripped Transport Canada in the 1990s, insisting that market forces would have to solve the ills of this most volatile service industry. Neither a clientelist nor a state-directed network could take root.

Though it did not follow the US model, air industry structure changed dramatically in Canada, and a duopoly soon emerged. In this phase of intense duopolistic competition, the capacities to articulate a common sectoral interest withered. Not by coincidence, the Ministry of Transport (MOT) implemented a series of bold privatizations and user-pay arrangements that transformed the infrastructural side of the industry. Arguably, the air sector has yet to settle into a corresponding network form. Given Air Canada's centrality, a dominant thread links Robert Milton and his government relations staff with the office of the minister of transport. However, the growing relevance of the Competition Bureau offers an alternative policy template that could—and many argue, should—overshadow the MOT, while the Department of Foreign Affairs and International Trade continues to hold a crucial role in defining international air prospects.

State Capacities Make a Difference

The outcome of conflicts in business politics are heavily influenced by the policy capacities of state actors. As mentioned earlier, state capacities are far from uniform. They can be expected to vary by jurisdiction, by industry, by instrument, and by time. If capacity connotes an ability to produce or achieve results, then it is important to be clear about just what results are being sought. States can be strong in certain dimensions and weak in others. Looking at the three industries under discussion, it could be argued that, as different as they were in character, the policy frameworks in place until approximately 1980 delivered positive results on the criteria of economic growth and profitability. This cannot be said broadly for the more recent policy frameworks in any of the three.

In Chapter Four we saw that, for half a century, the domestic air transport system was successfully moulded in Ottawa's preferred image, by a package of state regulatory and ownership levers. The business was overwhelmingly profitable, safe, and efficient in servicing a traffic base from coast to coast. The licensing of entry and exit, review of scheduling, fares, and revenue returns, and the acceptance of cost recovery and cross-subsidy principles, all furthered the fine-tuning of the commercial air system. A layered system of local, regional, and national service was the result. In addition, the state provision of infrastructure underwrote the operating air service business in crucial ways. This policy system placed a premium on workable clientelism between carriers and regulators, and the general acceptance of the underlying principles meant that tactical politics concentrated on adjustments at the margins. While the number of political stressors mounted after 1980, its resilience as a business system was not in doubt.

It was the shift of political balance between regulatory and counter-regulatory coalitions that offered the opportunity for air policy change toward domestic open entry, fare and schedule competition, and liberalized intercorporate merger and acquisition opportunities. This was reinforced by the simultaneous privatization of federal and provincial government airline holdings. Not all regulatory measures were abandoned, but the rules of the air carrier business were transformed about as dramatically as it is possible to imagine. In key respects the new consumer interest goals of reduced fares and flexible service were achieved. What was lost, however, was the capacity to shape the evolving business structure or the underlying financial stability of the industry. The federal government found itself without the tools either to guide this transition or to restrain private corporate tendencies. Even more significant, the rather sudden abandonment of direct regulatory instruments was not cushioned by an alternative policy framework of competition oversight or infrastructure regulation. Indeed, Ottawa had no political response for the ills of the new airline duopoly except to call for a "market-driven solution." Ministers repeatedly announced the absence of remedial tools and their unwillingness to contemplate measures toward re-regulation. In 1992-94 and again in 1999, the minimal reactive postures of ministers of transport is striking testimony to the shift, and loss, of policy capacity.

In dealing with steel, Canadian governments experienced a different transition in policy capacities that was only slightly less dramatic. During the high growth years from 1950-75, Ottawa's policy-strategy consensus proved to be a durable and effective, if politically informal, understanding between state and firms. Leading producers focussed upon the domestic market, concentrated on products of competitive advantage, planned output levels at average rather than peak demand, and relied upon imports to service the residual niches and the cyclical peak demand.

The result was a set of world competitive but domestically oriented producers. This policy mix was only eroded by the rise of global oversupply and the heightened import sensitivities of domestic steelmakers in both the US and Canada. With the rising levels of Canadian steel export and import, the critical policy arena shifted to the contingent protection system of dumping and countervail. In both nations, primary steel was the most heavily engaged industry.

Not only did steelmakers face the challenge of learning the folkways of CITT and ITC deliberations, as mediated through the GATT/WTO and NAFTA treaties, it also became clear that the procedural and methodological practices of the import policing bureaucracies were poor sites for fashioning industry-specific solutions. The 2002 CITT steel safeguard case illustrates the mixed outcomes that routinely characterize the field.

So far as pulp and paper is concerned, the major policy tools may well have resided in provincial capitals in the shape of crown forest ownership, leasing and stumpage payment policy, and electrical energy provision. Unlike the other two cases, state equity ownership has not been a common technique for shaping the structure or performance of pulp and paper enterprise. Instead, provincial governments have concentrated on attracting inward investment by large, integrated mill operators, treating these manufacturing establishments as business anchors for wider chains of forest commerce. In Ottawa, the pulp and paper industry was similarly regarded as a leading productive sector with superior performance by employment, efficiency, and export-earning criteria. Indeed this historically privileged position threatened to pose policy problems as the age, scale, and technology base of Canadian mills began to change in the international division of labour.

These traditional understandings were altered by the rise of an alternative regulatory template, made all the more threatening by its location several steps removed from the lands, trade, and industry agencies. The new environmental protection paradigms that emerged in the 1970s elevated the problem of mill effluents, hitherto treated as an "externality" to industrial operations, into a policy value of the first order. Pollution control legislation had the potential to alter core commitments of corporate strategy and competitive advantage. As seen earlier, the story of pulp and paper effluent regulation is one of successive phases of tripartite encounters between industry, state, and ENGOs. This involves a succession of policy instruments, including the all-important grandfathering exemption of older mills in 1970, the urgent extension of standards to include dioxins and furans in the late 1980s, and the shift away from legislated regulation and toward voluntary compliance with third-party standards in the 1990s. Here the sharing of federal and provincial jurisdictions played a key mediating role, as it also did in the case of pulp and paper modernizations incentives.

Beware of Simple Dichotomies

Business politics does not usually take the shape of an "us-them" or "corporate-government" polarity. More common is a pattern of business-government alliance, based on shared interests or policy preferences, facing off against one or more rival business-government alliances rooted in opposing interests. Since such political alignments are seldom identical from one issue to another, the way is open for complicated situational variations.

Certainly in steel there have been multiple patterns of overlapping alliances. Historically there have been regional alliances pitting Scotia Steel and Besco and their Nova Scotia government counterpart against Stelco and Algoma and Ontario on matters of freight rates. More recently, the Sysco-Algoma rail dispute reflects a similar alignment. There are also business alliances forged along commercial chains, as with the cross-ownership or rent chain connections of Stelco or Dofasco with iron mines, coal properties, and limestone quarries. Two North American cooperative research ventures linking steel manufacturers and customer industries are the NewSteel alliance for product innovation and the Ultra-Light Steel Auto Body consortium.

Coalitions between capital and labour have become increasingly evident as well. The Canadian Steel Trade and Employment Congress was established by steel firms and the USWA to defend jobs and sales in domestic markets against "unfair" imports and to defend strategic export enclaves against predatory trade actions. Similarly in the US, integrated firms and steel unions have acted in concert with the Congressional Steel Caucus to gain leverage on the legislative front.

Among steel firms themselves the orientations can vary. Consider the difference between the US, where integrated steel and mini-mills are sufficiently at odds politically to require separate and rival trade associations, and Canada where, to date, one organization has effectively built a political coalition of the two distinct segments.

In pulp and paper, where both levels of government play potentially powerful roles, the associational landscape combines national and provincial bodies. The CPPA has long functioned as the voice of pulp and paper, with the capacity to monitor and deal with both jurisdictions. At the same time, pulp and paper firms tend to be active participants in provincial "forest products" associations where they join with logging and lumber producers to manage shared interests in crown forest policy, labour issues, and environmental concerns. This can be reinforced by commercial links of fibre exchange, in which pulp firms sell sawlog quality material to the lumbermen, who in turn sell edgings and chipped mill wastes to the pulp producers.

One of the most interesting alignments emerges on pulp and paper effluent regulations. Initially, the consensual business position was to delay compulsory limits in favour of voluntary targets and to favour provincial as opposed to federal mandates. This reflected the shared predicament of Canadian producers, regardless of technique or location, in financing the secondary treatment facilities necessary to retrofit their existing facilities. As time progressed, however, and the second wave of PPER rules emerged with their focus on the dioxin-chlorine nexus, the cleavages between older and newer, kraft and non-kraft, open loop and closed loop firms became politically operative. Moreover the more "green" segment of producers appreciated the competitive advantage that could be gained over their rivals by meeting eco-standards. In the third wave of firm-based third-party certification of sustainable standards, this can be taken further again.

When it comes to airlines, a number of fascinating political polarities can be discerned. We have seen that, despite their limited jurisdictional purchase, provincial governments have demonstrated a sharp awareness of the economic values of scheduled air service. Several provinces adopted regional carriers as favoured instruments over the years, including the investment in equity positions to ensure the survival of regional hub enterprise. Nationally, the special contractual relationship between TCA and the federal cabinet was perhaps the most graphic example of a privileged government-business alliance.

In the closing years of the regulated domestic industry, the air carriers acted in virtual concert to defend the CTC regime against the advocates of market competition. In this the airlines were joined by their unionized workers, as a half-dozen separate unions formed the Council of Unions in the Aviation Industry. On the opposing side were the consumers lobby together with the peak business associations who sensed an opportunity to reduce the costs of doing business.

However, the greatest business affinity factor in airlines involves, without doubt, the emerging international alliances among free-standing carriers. It was the Oneworld alliance with American Airlines that sustained Canadian Airlines International during the financial travails of the 1990s. Similarly, it is notable that United Airlines, facing bankruptcy in 2002, turned to its Star Alliance partners Lufthansa and Air Canada for financial support.

Never Discount the Wider Political Context

As important as the field of micropolitics can be, there is a broader politics whose tide washes across all business sectors. Shifts of metapolicy frames can alter the terms of political advantage quite starkly. This points to a level of politics that often operates beyond the direct control of industry-level actors. If they are disciplined and determined, governing parties can define and control a policy agenda of half a dozen big-ticket issues. The same can be said for the departmental level, where ministers and senior officials address narrower but still substantial mandates. In most cases, a set of underlying values and practices is woven into an overarching policy frame, and significant transformations to such frames can have profound effects.

Recall the distinction between foundational policy-making, which defines the underlying framework of action, and normal policy-making, which is concerned with the more routine issues that arise within the bounds of such frameworks. While most business politics is of the latter type, foundational shifts are even more profound. In the previous chapters we have seen several important examples.

During the Trudeau years, one signature metapolicy was the need to formulate strategies for economic growth in depressed Canadian regions. Indeed it was under this mandate that Ottawa's Department of Regional Economic Expansion was fashioned. The Atlantic provinces figured strongly in these plans, with several funds and agencies designed specifically to address regional needs. There were also sectoral planning processes that channelled joint government support to priority industries. One such target was the modernization of the Sydney Steel mill in Nova Scotia. Under the terms of federal-provincial agreements stretching over the better part of a decade, Sysco's production process was transformed from open hearth steelmaking to electric arc operations. Whatever the impact of these measures on the ground, the plan was entirely consistent with Ottawa's policy commitments, and the refit was accomplished. In so doing, however, it triggered several powerful political reactions. One came from Sault Ste. Marie, where Algoma Steel fought the principle of state aid for a weakened business rival. Another emanated from Washington, where US steel rail producers mounted countervail actions against Sysco and Algoma exports to the south. Finally, Sysco's "fit" with Ottawa's 1970s-style industrial policy thinking led to preferential procurement battles with the two national railways. A second outcome of this industrial-regional policy regime was the pulp and paper modernization agreement of 1979 that tackled the problem of aging capital assets in eastern Canadian mills.

In only a few short years, these intellectual and policy coordinates were transformed both in Ottawa and some provinces. This was driven by multiple factors: the shortcomings of fiscal Keynesianism in a stagflationary decade; the inexorably mounting budget deficits that limited fiscal capacity; the rapid demise of the 1980 National Energy Policy, Ottawa's last bold sectoral initiative of the postwar era; and the demonstration effect of neo-liberal economic policies in Thatcher's London and Reagan's Washington. The final Trudeau government began, reluctantly and haltingly, to undertake such a transition between 1980 and 1984. A cabinet that opened its mandate with bold programs for energy-driven megaprojects and heightened capital investment controls morphed, over four years, into the sponsorship of deregulating service industries (from transport to communications to finance) and renovating the competition regime. Axworthy's open-entry gambit with the Air Transport Committee was a case in point.

Yet it was left to the Mulroney Conservatives to complete the process. By mid-decade, the government had developed a new economic metapolicy whose centrepiece was the strategy of comprehensive free trade with the US. This arose at a time when Washington was aggressively pressing its trade partners to conform to a variety of "fair trade" practices rooted in US law. As the draft agreement firmed up, a technically complex and politically volatile debate erupted in Canada over the implications this held for the national interest. Yet as far as steel was concerned, a central element in import policy was the opportunity for domestic firms to petition for punitive duties on incoming goods that might have "benefited" from government subsidies. This was the nub of the steel rail dispute and of a wider set of defensive manoeuvres by the beleaguered integrated steel sector. In effect, the advent of Canada-US free trade forced a wide range of Canadian economic policies to be reviewed for their trade disruption potential, with many being altered and even dismantled as a result. While this policy shift was rooted in forces far broader and deeper than the steel industry alone, it had profound effects upon the politics of that sector.

The conduct of the Liberal government after 1993 brought its own signature metapolicies. First among these was the drive for fiscal control in the form of a balanced budget. From 1994-97, the Chrétien-Martin "program review" touched virtually all areas of public administration. It could hardly be otherwise, when the challenge was to close a deficit of $35-$40b that had defied government efforts for two prior terms. Of the industries explored above, it was airlines that felt the budget knife directly. Transport Canada was dramatically transformed across all modal areas. Airports were transferred to local authorities, air traffic services were privatized, and an array of surcharges and user fees were levied in efforts at cost recovery. All of these had a direct impact on the economics of air carriage, as Air Canada Tango revealed when its special "promotional" fare of $2 for a coast-to-coast ticket was hiked to $97 by surcharges.

Ottawa's Green Plan was another metapolicy whose rise and decline affected an array of environmental initiatives. First introduced by the Conservatives in the late 1980s, it outlined spending and regulatory plans on an ambitious order. It also levered environmental issues up the federal policy agenda, including measures affecting pulp and paper effluent control. While the funding levels for Green Plan initiatives were compressed in magnitude and extended in roll-out, the window of opportunity from 1988-92 was instrumental in the redefinition of emission standards and the realignment of industry-public positions.

A parallel syndrome can be detected in commercial air service. The postwar framework was based on the acknowledged need to manage a stable and safe air travel network. To this end the state regulated the entry, ownership, routes, and fares of all carriers. In the 1980s, a new theme emerged to serve different goals. Now the public interest was defined in terms of the need to reduce the (inflated) cost structures of a sheltered industry and to reap consumer benefits.

Finally in the case of pulp and paper, the postwar framework aimed to facilitate the profitability of a model traded commodity industry. Not only was it based upon a world-class natural resource endowment, but it also led all manufacturing in full-time employment, and it figured as the leading export earner of foreign exchange. While these factors continued to apply, the policy context was transformed significantly in the 1970s and beyond. New scientific findings on the toxicity of pulp plant effluents brought an overarching concern for the protection of public health and ecological integrity.

A Final Word

If one thing is certain, it is that the future will bring changes to the strategies, structures, and policies discussed in the chapters above. This is to be expected and welcomed. It offers an opportunity to weigh and adjust the interpretations developed here against the ongoing development paths in paper, steel, and airlines. In the postscript that follows, several of the leading political issues are reconsidered in light of developments in the year 2003.

Micropolitics is as fascinating as it is complex. The more light that can be shed on its possibilities and variations, the more we can ultimately understand about the intriguing figure with the greatcoat, walking stick, and bowler hat.

CHAPTER SIX

Postscript:
Micropolitics
Marches On

A striking theme, which permeates all of the chapters above, has been the need to recognize the complex and deep-seated forces that can drive political change within firms and industry groups. This reality presents both problems and opportunities for a book of this sort. One problem stems from the desire to keep current. Decades ago, British Prime Minister Harold Wilson remarked insightfully that a week is a long time in politics. With apologies to Wilson, we might say that a year can be a lifetime in the micropolitics of business. The Chrétien era of federal leadership gave way, in December 2003, to that of Paul Martin. The Canadian dollar, after hovering in the US$ 0.65 range for several years, rose as high as US$ 0.78 during 2003.

This final chapter tackles some of the dramatic new developments that have arisen since the manuscript was finalized. Their inclusion offers one last stab at the understanding of these three industries and a final opportunity to explore the links between conceptual frameworks and case material.

For pulp and paper, the turn-of-century trend for mega-corporate mergers continued its course. In November 2003, west-coast Canadian giants Canfor and Slocan announced their intention to merge through a share swap deal.

For steel, the global trade situation took another significant turn, with the WTO's ruling against the Bush steel safeguard package of 2002. This challenge, and the ensuing reversal of fortune, carry significant implications for Canada as well. It offers another example of policy linkage and of the conditioning role of international economic regimes in constraining the exercise of sovereign state policy. The steel sector was also shaken by Stelco's move to file for bankruptcy protection, early in 2004.

For airlines, the single most important development of 2003 was Air Canada's decision to seek bankruptcy protection. This highlights the significance of politics within a single firm, as the company struggles to keep afloat.

The Softwood Lumber Dispute and the Canfor-Slocan Merger

The west coast pulp and paper industry is closely aligned with lumber production, with most leading firms combining the two operations. Consequently, the
revival of the Canada-US softwood lumber dispute in 2001 impacted quickly
on pulp and paper prospects. More importantly, the failure of the two nations
to reach an early settlement raised the spectre of long-term loss of Canadian
access to the American market. This formed the backdrop to the dramatic
announcement on 25 November 2003 of a forest sector mega-merger. Two of
British Columbia's largest firms, Canfor and Slocan Forest Products, agreed
to combine.

The resulting giant is the largest spruce-pine-fir lumber manufacturer in
the world and the second largest lumber manufacturer of any kind in North
America. The "new" Canfor's 2004 capacity exceeds 5b board feet in lumber,
1.2m tonnes of pulp, 950m square feet of plywood and oriented strand (chip)
board, and 150,000 tonnes of kraft paper.

The trans-border lumber dispute is a powerful contextual force for this
consolidation. In April 2001, following the expiry of the previous Canada-US
agreement on softwood exports, US producers resumed their tactics of attack-
ing Canadian exports on anti-dumping and countervail grounds. At the core
of the dispute was the Canadian practice of allocating crown timber according
to stumpage agreement rather than commercial auction. Large US sawmill
operators in both the northwest and southeast states contended that the stumpage system represented a public subsidy and an unfair competitive advantage
for the Canadians. This has been a continuing refrain since the late 1980s. In
the fall of 2001, Washington responded to the complaint petitions by levying
preliminary anti-dumping and countervail duties of 27 per cent. This was
confirmed the following year, when final duties were levied in almost the same
amount. The commercial impact of these trade penalties is severe. It obliges
producers in British Columbia to pay a significant surcharge to the US government on every export shipment. Slocan, for example, lost almost $58m from its
net sales in 2002. This was considerably more than its net income of $37m.

Since much British Columbia pulp is destined for Asian rather than US
markets, the US softwood duties do not directly affect this sector. However,
the rebound effect from depressed lumber receipts, together with sharply
reduced overall profits, is compelling. In addition, there is no evidence to suggest an early negotiated settlement that will lift the penalties. Despite repeated
talks, Canada appears to be far more eager than the US to come to terms. The
looming prospect of a presidential election campaign in 2004 may further nar-

row the negotiating window. Moreover, at home the government of British Columbia is in the process of revising its crown timber policy, reducing the allocations to licensees, and moving toward an auction system.

In short, almost all of the basic forest industry coordinates of timber supply and lead market access are in flux. Having wrung all of the plant-base efficiencies out of their respective mills, Canfor and Slocan managers decided that the next level of performance gains must be sought from corporate restructuring on a grand scale. It remains to be seen whether this marks the start of a new wave of consolidations across the western industry.

North American Steel Safeguards and the WTO Challenge

On 10 November 2003, a WTO Appeal Panel confirmed the finding that the US safeguards package announced by President Bush in March 2002 was inconsistent with the terms of both the WTO *Agreement on Safeguards* and the WTO founding Agreement. This ruling required either that Washington eliminate the safeguard measures or the complainant states, led by the European Union and Japan, would be free to retaliate by imposing duties worth billions of dollars on incoming US goods.

Such cases can be understood on at least three levels: as a technical trade dispute being arbitrated against the terms of treaty economics and law; as a policy tool adopted by the host, or importing, state as a result of a complex domestic political process; and as a step in the broader politics of the global steel battle, in which nation states and national industries manoeuvre for competitive advantage.

First, on the surface, it is a technical dispute over permissible actions under an international commercial trading regime. The WTO system authorizes member states to take "safeguard" actions to stabilize their domestic markets in the face of serious injury due to sudden and unexpected surges in imports. The detailed terms of such interventions are regulated by the WTO *Agreement on Safeguards*, a 14-article deal that regulates activities under WTO Article XIX, which covers emergency action on imports. The thrust of these rules is to permit short-term import controls, by duties or quotas, provided that the host state complies with terms and procedures set in the agreement. For example, safeguard measures must be applied only after the imports have been determined to cause serious injury, must be applied to all import flows regardless of source, must be limited to a four-year maximum period, and must be determined only after notification of WTO authorities.

The steel-exporting states that stood to be harmed by the Bush safeguard package lost little time in launching challenges at the WTO. By August 2002, the European Union had been joined by Japan, Korea, China, Norway, Switzerland, New Zealand, and Brazil. The WTO panel concluded, just under a year later, that Bush's ten safeguard measures did not square with the agreement. While the details varied by product group, the underlying flaws were found to be the US failure to show sufficient causal links of imports to serious injury, the inconsistent and alternative explanations which the Americans advanced for different product groups, and the failure to achieve parallelism in applying US safeguard measures to all import sources within product groups.

The US then exercised its right to appeal several issues of law and interpretation by the WTO panel. This led to a second set of appeal hearings in September 2003 in which the US asked that the July ruling be set aside. Washington contended that the panel erred by adopting several evaluative criteria that are not explicitly sanctioned in the agreement. It challenged, for example, whether the panel was authorized to weigh the adequacy of the ITC explanation, arguing instead that, so long as the ITC analysis was reasoned in a detailed analytical form, it was in compliance with the agreement. In effect, the US argued that the panel was writing new rules into the agreement by the terms of its decision.

In November, the WTO appeal panel upheld the initial ruling and instructed the US to achieve compliance (Morton 2003). On 4 December 2003, President Bush announced the lifting of the safeguard measures, stating that their goals had been achieved and that any further continuance would impose excessive hardship on US steel-consuming industries.

This White House rationale was clearly one of convenience rather than substance, since in the absence of the WTO ruling the safeguard measures would likely have run the remaining 16 months of their intended three-year course. However, it does highlight the deep currents of domestic politics that swirl around any trade policy measure. As earlier chapters revealed, the imposition of tariff or quota protection for one industry serves to raise the cost structures for that industry's consumers. In this case, the cleavage ran between industries and also between lobby groups. The steel producer alliance with steel-sector labour is expressed through the US "New Steel" campaign, while the counter-pressure has been applied by the auto industry and other steel users. At the same time, a curious battle of competing experts broke out on the question of the distributive impact of the safeguard measures. The American Iron and Steel Association commissioned one set of consultants to demonstrate a relatively minor impact on downstream steel users while the

Consuming Industries Trade Action Committee marshalled quite the opposite.

Steel Safeguards and Canada

Inevitably, the WTO steel politics had a powerful impact north of the forty-ninth parallel. Earlier it was shown how the Bush safeguard package interacted with Canada's own steel safeguard deliberations. First of all, the ITC findings of 2001 prepared the way for Canadian steelmakers to call attention to their own offshore injuries. Second, the decisive breadth and depth of the Bush tariff measures of 2002 offered a standard of comparison for the Canadian industry's lobby of Ottawa. Not only had an economically liberal president embraced defensive trade measures but, more particularly, the exemption of NAFTA steel industries from the US safeguards program represented a vital feature that Canadian producers and unions sought to reciprocate in their case brought before the CITT in 2002. The two steel sectors were pursuing strikingly similar import policy agendas.

Equally important in Canada, though, was the impact of the WTO proceeding against the Bush safeguards. Since the outlines of the European and Japanese cases against Washington were evident months before the CITT report of August 2002, the Canadian agency had a dramatic foreshadowing of the global steel reaction to a Bush-style ruling. The same held for the broader international trade bureaucracy in Ottawa. At the very least, the WTO steel challenge had to sensitize the CITT to the impending close scrutiny of *its* evidence, findings, and recommendations. The sequence of key US, world, and Canadian steel safeguard actions is portrayed in Figure 6-1.

There were likely many causes for the CITT's rather tepid safeguard proposals. But their far less sweeping scope, together with the denial of a US export exemption, served at the very least to shield Canada from the sort of challenge being raised against Washington. In Ottawa, a political stand-off appeared to follow. Steel industry and labour interests excoriated the CITT and lobbied the Chrétien government for a more sweeping final package. On the other hand, steel importers and consuming industries followed the lines already well rehearsed by their counterparts in Washington.

For more than a year, the cabinet deferred a public response of any kind. Finance Minister John Manley played for time in December 2002 by appointing a Steel Working Group drawn from the federal Liberal caucus and the CSPA. That group's report, four months later, reprised the case for bold safeguard action. Newspaper reports in the spring of 2003 described the

Figure 6-1

Steel Safeguard Politics, 2002-03

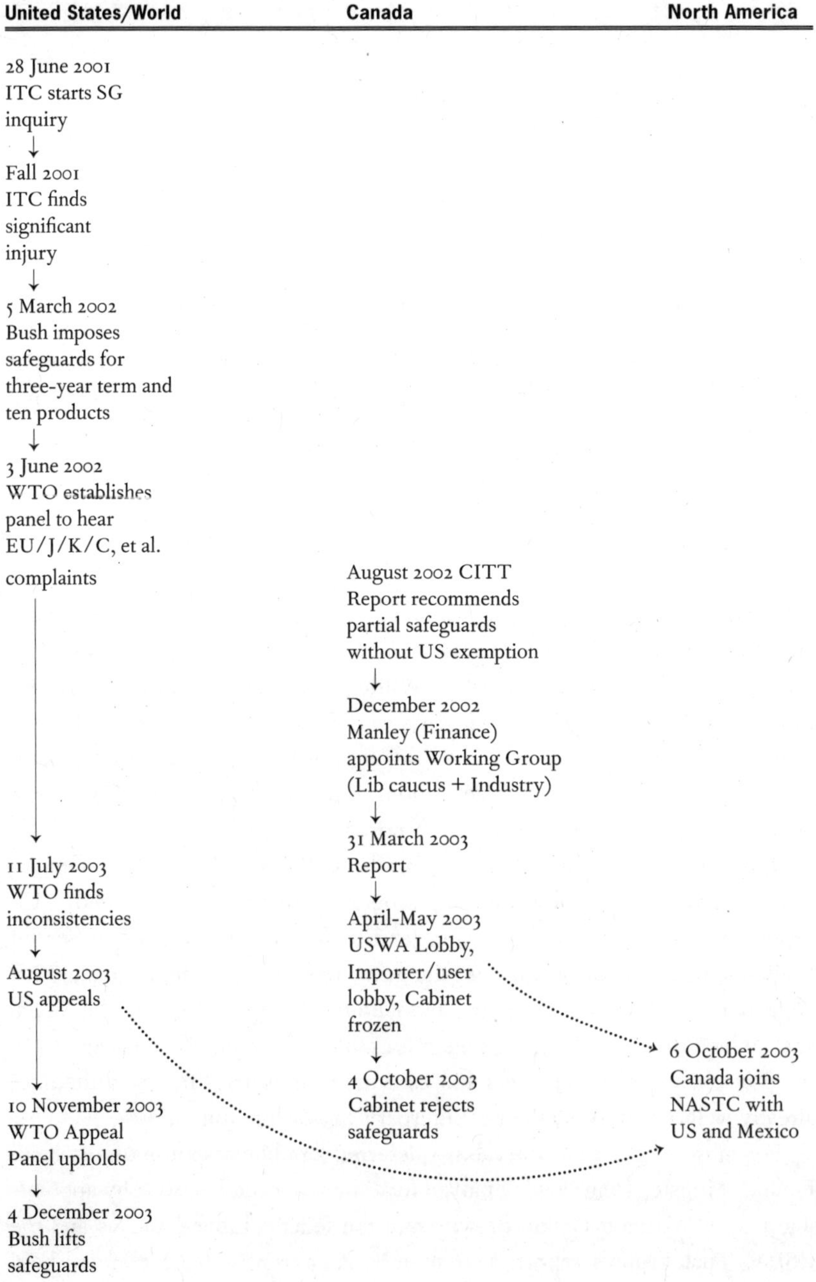

United States/World	Canada	North America
28 June 2001 ITC starts SG inquiry ↓ Fall 2001 ITC finds significant injury ↓ 5 March 2002 Bush imposes safeguards for three-year term and ten products ↓ 3 June 2002 WTO establishes panel to hear EU/J/K/C, et al. complaints		
	August 2002 CITT Report recommends partial safeguards without US exemption ↓ December 2002 Manley (Finance) appoints Working Group (Lib caucus + Industry) ↓ 31 March 2003 Report ↓	
11 July 2003 WTO finds inconsistencies ↓ August 2003 US appeals	April-May 2003 USWA Lobby, Importer/user lobby, Cabinet frozen ↓	
↓ 10 November 2003 WTO Appeal Panel upholds ↓ 4 December 2003 Bush lifts safeguards	4 October 2003 Cabinet rejects safeguards	6 October 2003 Canada joins NASTC with US and Mexico

continuing stalemate, along with compromise safeguard measures that were considered at the cabinet level but then rejected in the face of deep-seated ministerial disagreement (Brieger 2003).

For the anti-safeguard coalition, the extended political delay was a tactical victory in two respects. Not only was the political momentum for safeguard action slowed to a halt, but, equally, the impending release of the WTO panel report warned of an additional political reality check. In the Ottawa bureaucracy, the consensus strategy became one of patience and reconsideration. Once the WTO ruled against the Bush safeguard package, the political sentiment in Ottawa seemed to follow suit. In many ways it was an overdue formality when the cabinet announced, on 4 October 2003, that no steel safeguard actions would be taken in Canada.

This position was linked to signs of an alternative continental strategy of Canada-US-Mexico collaboration to pursue a common steel trade reform agenda. A new body, known as the North American Steel Trade Commission, held its inaugural meeting in November 2003, pledging to work with industry to coordinate positions for future OECD steel subsidy talks and to reduce distortions in the NAFTA steel bloc with the eventual goal of a free trade bloc.

Air Canada and the Special World of Bankruptcy Protection

After 1999, as it struggled to absorb the complicated pieces of CAI, Air Canada management also fashioned a new business strategy for an era of single-firm dominance. Described in Chapter 4 above, this involved a package of subsidiary carriers serving niche markets under the authority of a parent holding company. Not only would these carriers compete with corporate rivals such as Westjet, Jetsgo, Skyservice, and (before its demise) Canada 3000. Equally significant, they would, over time, compete with one another. More precisely, there was clear potential for customer traffic to switch between subsidiaries and for the commercial efficiencies of the more successful to penetrate across the Air Canada domain.

However, the multi-carrier model was put into question by Air Canada's relatively sudden financial failure, as signified by its 1 April 2003 filing for bankruptcy protection under the Corporate Creditors Arrangements Act. This new crisis was the culmination of pressures in the winter of 2002-03, when a new string of difficulties depressed global airlines. The second Iraq War had the dual effect of driving up fuel costs and dampening discretionary travel. Almost simultaneous with the onset of hostilities came the discovery of the SARS epidemic. Not only did this cut in half the level of Asian air travel but, given Toronto's epicentral status for SARS in North America, it posed a special threat to Air Canada's hub traffic. The *coup de grâce* came when Air Canada's unions resisted management's February call for a 20 per cent negotiated cut on labour costs, together with sustained pressure from the federal pension regulator that the company address gaping arrears in employee pension obligations.

As seen earlier, Air Canada was not the first airline to seek the protection of the bankruptcy courts in order to restructure its operations and finances. In the US, chapter 11 protection has played a major role in both the airline and steel industries. It is, in effect, the only practical alternative to business failure and dismantlement. In Canada, the Corporate Creditors Arrangements Act serves a similar role. Under the supervision of a judge and a court-appointed business monitor, the claims of creditors are effectively frozen for a period of time. This provides the company management an opportunity to cut costs, refinance, and generate a new business plan.

Bankruptcy protection, however, should not be considered a mere opportunistic time-out from normal business disciplines. It comes with a heavy price to corporate autonomy, for several reasons. First, the process triggers the insertion of new and potentially powerful external interests into the heart of the beleaguered firm. One is the judge designated to supervise the reorganization and, ultimately, to approve the results. Another is the outside monitor, normally an accounting or management consultancy, that helps shape the judge's findings.

Second, the predicament of bankruptcy can trigger a profound realignment among the constituent interests within the firm, quite apart from the new supervisory regime. Senior managers and directors may not survive the transition and, if they do, their freedom of manoeuvre can be severely curtailed. Nor will internal "stakeholders" escape unscathed. Workers will be required to make major concessions on wages, lay-offs, benefits, and work rules. Shareholders will likely see the value of their holdings slip toward zero due to dilution from new issues and the collapse of secondary trading. The sheltered company's creditors are an especially volatile brew, as they will be asked to settle their accounts for a fraction of face value. In effect, most parties to the beleaguered firm's "political coalition" will be asked to choose between hurt — partial economic losses as the price of continuance — and worse hurt — total economic losses in business failure. There is endless potential for hostility and in-fighting over the distribution of relative pain. This can turn the restructuring campaign into a battle of sequential vetoes, since any major interest bloc holds the potential to kill the recovery package. For Air Canada, these relationships are captured in Figure 6-2.

Figure 6-2

Internal Politics of Air Canada in Bankruptcy Protection, 2003

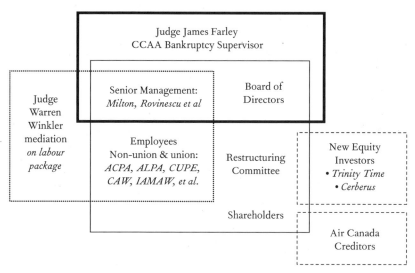

Finally, there is an expensive corollary cost of bankruptcy protection for any major corporate player. So consuming is the struggle to build the new business coalition behind the restructuring project that it overshadows all other considerations. In effect, it displaces a panoply of other pressing issues of business and political strategy that would normally be pivotal to competitive advantage.

This is certainly evident in the Air Canada situation. The first two months after the April 2003 bankruptcy filing involved the dirty work. With a reported $414m cash on hand and expenditures of $5m per day, Air Canada was approaching failure at terminal velocity. The board of directors turned down a federal government offer of a $300m loan, with its inevitable strings, in favour of a $700m infusion from private lender GE Capital. Simultaneously, company management sought new capital in any available direction: the Aeroplan contract with the Canadian Imperial Bank of Commerce was renegotiated, and a new American Express Aeroplan deal was signed.

However, the most difficult of these first steps toward survival involved a new deal on labour costs, which account for close to 40 per cent of FSNC budgets. Recognizing the continuing deadlock, Judge Farley took the unusual step of enlisting a judicial colleague, Warren Winkler, as a mediator. A former labour lawyer, Judge Winkler began an intense 18 days of talks on 13

May 2003. His first step was to summon the top Air Canada executives to an immediate meeting in Toronto. By all accounts, his relaxed but determined initiative was crucial to breaking the gordian knot (MacArthur and Willis 2003). Winkler instructed Air Canada to pare its concession demands to the core. He also persuaded Robert Milton to drop any proposed pension cuts from that package. Winkler was then able to bring the unions to the table to agree to what one newspaper described as the largest set of labour concessions ever granted to a Canadian company: $766m per year on a $3b labour budget.

With nine bargaining units at the table, this was far from straightforward. Yet when, ultimately, only the Air Canada Pilots Association was unwilling to sign on, Farley announced that he would hear arguments on the company's liquidation on 1 June. In one last series of meetings, Winkler fashioned a compromise between the mainline and regional pilot groups, each represented by a separate union, over work rights to the next generation of small (under 75 passenger) jets. This completed the deal and preempted the liquidation hearing.

Only with the labour package in place was it possible to invite new equity proposals. In the second phase of the restructuring, from June to November 2003, Air Canada pared a long list of potential investors down to a short list of two. Each of these combined an injection of new capital (in return for voting rights and seats at the board) and a new share rights issue. The latter enabled Air Canada creditors to purchase equity at the same price as the new partner, converting unrecoverable debt into shareholdings in airlines.

One of these bids came from Victor Li through Trinity Time Investments Inc. The son of Hong Kong Canadian tycoon Li Ka-Shing, Victor was closely involved with the family's successful takeover and turnaround of Husky Oil. Trinity offered to buy 31 per cent of Air Canada's shares for $650m, with Deutschebank underwriting $450m of share rights for creditors. Another tangible consideration was that Victor Li, as a Canadian citizen, was unaffected by the statutory 25 per cent limit on foreign shareholdings.

The second finalist was Cerberus Capital Management LP of New York. Named after the mythical three-headed dog that guards the gates of Hades, Cerberus is a hedge fund that specializes in the turnaround of failing firms. It offered to invest $500m in new equity along with a $500m rights issue.

Shortly after the Air Canada board announced on 8 November that it would recommend the Trinity bid to Judge Farley, this fragile political coalition was shaken by a new challenge. Cerberus announced that its previous offer was not its best bid. It then revealed not one, but two enhanced bids: one for a 27 per cent share with $650m equity and another for a 12 per cent share with $250m along with a rights issue of $850m (Waldie 2003). The latter was aimed at the interests of Air Canada's increasingly restive creditors, allowing them a larger potential recovery on losses. However, this follow-up offer also raised implicit questions about the sincerity of the initial proposal. Not surprisingly, the Air Canada board rejected any consideration of revised bids and defended its choice of Trinity as the best rounded package. In a follow-up court hearing, Cerberus failed to win an order obliging the Air Canada board to consider the follow-ups, and just before Christmas, the Trinity offer prevailed. Air Canada entered 2004 with a new controlling shareholder.

Or so it seemed. Almost immediately, Air Canada suggested that the huge employee concession package of 2003 would have to be broadened in order to meet the company's $1.1b target. Then Trinity pressed for a basic redesign of the Air Canada pension plan in a direction that shifted future risk from the employer to employees. In the face of union resistance, Trinity walked away in April and the business press began speculating on an Air Canada liquidation. Once more on the brink, one final round of intra-firm negotiations was launched. Deutsche Bank, in association with the major creditors, agreed to almost double its equity (covering much of Trinity's former share). GE Capital agreed to maintain its $1.8b loan package for an extended bankruptcy protection period. Finally, at the eleventh hour, the unions reached a new compromise package on labour costs. Together, these terms gave a measure of new life to Air Canada's battle with bankruptcy.

It is interesting to note that these developments unfolded in an open-ended political process, made possible by the rules governing bankruptcy protection. Equally notable is the fact that the political domain of corporate refinancing and reorganization is completely separate from that of federal policy and its political network. (These points also apply to Stelco, in its struggle with the burdens of the bankruptcy protection granted in January 2004. Interestingly, Judge Farley presided over this case as well.)

Conclusion

Some of the commentary in this book begins to age as soon as the print dries. More lasting, however, are the frameworks, concepts, and interpretations on which the discussion is based. The forward march of politics in our three industries offers an opportunity to reconsider these fundamentals. How durable are the insights and how portable are the explanations? Even as the forms and patterns and policy problems evolve, do these chapters offer a valuable foundation for tracking new developments? Readers will recall the argument, advanced early in this study, that a flexible set of analytic perspectives can travel widely in the world of business politics. The perpetual motion of rapidly changing events can always be profitably weighed against the slower dynamic of underlying structures.

Micropolitics on the Worldwide Web

Internet resources can play an indispensable role in the study of the micropolitics of business. However, it is important to put web resources in the proper perspective. They cannot substitute for the staple materials of scholarly books and articles, government publications and annual reports, journals, periodicals, and newspapers. The value of the web lies in supplementing these materials and in extending your reach through time and space. Almost always, internet searches will prove more meaningful and rewarding once you have become familiar with the fundamentals of the firms, industries, and public policies of interest.

It is quite unwise to begin your research project by logging onto the web. Too often this can result in a jumble of fragments and factoids of varying relevance and authenticity. Published materials still offer the fastest and most efficient orientation to a new subject. In most cases, far more can be gained from three hours immersed in well-chosen published works than from the equivalent time spent surfing on the web. However, once it becomes clear what you are seeking, the internet comes into its own: filling gaps, identifying specialized topics, updating materials, and extending your reach.

Here are some suggestions for pursuing online research. First, be flexible in using internet search engines. Try a variety of key search words if the first and most obvious fail to work. Second, pay particular attention to the onward links that are featured on many sites. Be systematic in exploring these links, keeping track of your progress to avoid getting lost in the maze. Print out valuable findings at frequent intervals. Third, be systematic in noting your sources, treating each internet address (URL) as a documentary source in the same way as you would an article or book or report. It is crucial to cite your internet sources in your work by title, author, URL, and date when accessed. Fourth, recognize that internet resources are part of a special public domain, and approach them accordingly. The by-word must be "researcher beware." Recognize that URL sites have been established for particular purposes: promotion, advocacy, marketing, and others. As such, these sites need not be balanced, fully accurate, or comprehensive. Accept this as both a strength and a limitation of internet materials and you can add an invaluable dimension to the understanding of micropolitics.

The following sections offer an introduction to some useful internet sources related to the chapters in this book. Bear in mind that such sources—and their URL addresses —are changing constantly and expanding in scope. This section was compiled in 2004.

Pulp and Paper

All of the leading forest products firms maintain corporate web pages, including Domtar Inc. <www.domtar.com> and Stora Enso <www.storaenso.com>. A wide range of corporate contacts can be found on the Roper site <www.canadian-forests.com>. The Forest Industry Network <www.forestind.com> describes itself as a "virtual community" on the internet.

Among trade associations, the Forest Products Association of Canada (formerly the CPPA) is one of the largest and most powerful in the country <www.fpac.ca>. The US counterpart is the American Forest and Paper Association <www.afandpa.org>. In Canada, there is also a set of provincial-level forest products associations, such as the Quebec Forest Industries Association, <www.aifq.qc.ca>. In British Columbia, the Council of Forest Industries, <www.cofi.org>, has played a leading role in forging joint industry positions. Such groups draw together business interests in the logging, saw milling, and pulp and paper segments, with particular interest in crown timber leasing policy and operational issues that fall under provincial jurisdiction. An extensive list of forest-sector association links, both business and professional, can be found at the very helpful Roper site mentioned above.

At times alliances and coalitions dedicated to promoting or opposing particular policy issues appear. One such group is the Canadian Sustainable Forestry Certification Coalition <www.sfms.com>. Established to promote systems of third-party certification of sustainable forest management, its membership closely mirrors that of the Forest Products Association of Canada. A different approach to industry-government research collaboration is found in the Pulp and Paper Research Institute of Canada, or Paprican <www.paprican.ca>. The Canadian Institute of Foresters <www.cif-ifc.org> is a federated group representing professional foresters.

The trade press offers a variety of web contacts. *Pulp and Paper Canada* is the premier journal in this country <www.pulpandpapercanada.com>. An equally useful publication is *Pulp and Paper International*. Limited access to that journal and to its sister publication, *Pulp and Paper*, is available at Paperloop.com <www.pponline.com>. The same applies to Paper Industry <www.paperindustrymag.com> and Canadian Forest Industries <www.forestcommunications.com>.

As far as Canada's federal state system is concerned, jurisdiction over the forest industries is widely dispersed. In the government of Canada, timber policy matters are handled by the Canadian Forestry Service within Natural Resources Canada <www.nrcan.gc.ca/cfs-scf>. Pulp and paper manufacturing policy is centred in Industry Canada <www.ic.gc.ca>. Of special interest is the sector framework for pulp and paper <strategis.ic.gc.ca/SSG/fbo1002e.html>. Trade issues, on the other hand, tend to be concentrated in the Department of Foreign Affairs and International Trade <www.dfait-maeci.gc.ca>. When it comes to regulating pollution from mills, the key federal players are Environment Canada <www.ec.gc.ca>, and the Department of Fisheries and Oceans <www.dfo-mpo.gc.ca>. For pollution regulation in the US, the Environmental Protection Agency is the key national player <www.epa.gov/ostwater/pulppaper>.

Provincial governments exercise their rights to crown lands and forests title through departments of forests or natural resources. The leading jurisdictions are British Columbia <www.gov.bc.ca/for>, Ontario <www.mnr.gov.on.ca/MNR/forests>, Quebec <www.mrn.gouv.qc.ca>, and New Brunswick <www.gnb.ca/0078/index-e.asp>.

There are two large, nationally based trade unions in the forest products sector. One is the Industrial, Wood and Allied Workers of Canada <www.iwa.ca>. Based on the west coast but active in seven provinces, it concentrates on the primary forest sector (i.e., tree harvesting and hauling). The mill sector is also densely organized, by the Communications, Energy and Paperworkers Union of Canada <www.cep.ca>. The Pulp, Paper and Woodworkers of Canada is a British Columbia union affiliated with the Confederation of Canadian Unions <www.ppwc.bc.ca>.

A wide range of ENGOs is active in forest industry politics. These include, in no particular order, the Canadian Wildlife Federation <www.cwf-fcf.org>, the Canadian Nature Federation <www.cnf.ca>, Greenpeace <www.greenpeace.org>, the Canadian Parks and Wilderness Society <www.cpaws.org>, the Sierra Club of Canada <www.sierraclub.ca>, the Sierra Legal Defence Fund <www.sierralegal.org>, the World Wildlife Fund <www.wwfcanada.org>, the Nature Conservancy of Canada < www.natureconservancy.ca>, the David Suzuki Foundation <www.davidsuzuki.org>, and Friends of the Earth <www.foecanada.org>. A different and ENGO-critical perspective is offered by Patrick Moore <www.greenspirit.com>.

There are many research institutes and think tanks dealing with pulp and paper issues. They include the Finnish Forest Research Institute <www.metla.fi> and the International Union of Forest Research Organizations <iufro.boku.ac.at>. For professional forester opinion, the national associations are the Canadian Institute of Forestry <www.cif-ifc.org> and the Society of American Foresters <www.saf.org>.

Steel

Corporate pages for the major integrated steel companies include Stelco <www.stelco.com>, Dofasco <www.dofasco.ca>, and Algoma <www.algoma.com>. For the mini-mill segment, there is IPSCO <www.ipsco.com>, the Gerdau family of companies — Co-Steel, Gerdau-Courtice and Gerdau-MRM — <www.ameristeel.com>, and Ispat-Sidbec <www.ispatinland.com>. Specialty and stainless producer Atlas <www.slater.com> rounds out the list. Useful overviews of the Canadian primary steel industry can be found from the Industry Canada data base <strategis.ic.gc.ca/ SSG/mmo1070e.html>. For the international industry, try the OECD Steel Committee <www.oecd.org> and the World Steel Dynamics site at Paine Webber <www.steel.com/wsd>.

The central trade grouping is the CSPA <www.canadiansteel.ca>. Other associations include the joint business and labour Canadian Steel Trade and Employment Congress <www.cstec.ca> and the Canadian Institute of Steel Construction <www.cisc-icca.ca> representing the structural steel and platework fabricating industries. In the US, the oldest industry group is the American Iron and Steel Institute <www.steel.org>. The electric arc producers have their own group, the Steel Manufacturers Association <www.steelnet.org>. Many Canadian producers are active in both of these, along with the Specialty Steel Industry of North America <www.ssina.org>. Also pertinent is the Steel Recycling Institute <www.recycle-steel.org>. An interesting recent development has been the formation of The Steel Alliance <www.thenewsteel.org>, a grouping of some 75 North American steel producers who have united to promote and defend the renewed use of steel in core products such as automobiles, cans, and construction materials. See also the Ultra-Light Steel Auto Body alliance <www.ulsab.com>. There are also important voices for steel importer and user interests. The American Institute of International Steel <www.aiis.org> is a coalition resisting defensive steel trade actions and defending the position of traders and buyers. The Consuming Industries Trade Action Coalition <www.citac.info> is a cross-industry coalition that includes steel among its leading sectoral concerns. For global business trade matters, the International Iron and Steel Institute <www.worldsteel.org> plays a lead role. On matters of steel trade policy, the OECD <www.oecd.org> publishes annual industry reviews and more specialized studies. It also hosts and documents ongoing series of high-level meetings on steel problems.

When it comes to state agencies, there are a number of important sites within the government of Canada. As mentioned above, Industry Canada maintains a steel section in its manufacturing industries branch. For trade issues, the key offices for dumping and countervail are the Canada Customs and Revenue Agency <www.ccra-adrc.gc.ca> and the CITT <www.citt.gc.ca>. The Export and Import Controls Bureau of the Department of Foreign Affairs and International Trade administers the steel import monitoring program <www.dfait-maeci.gc.ca/~eibc>. At the provincial level, labour relations and airborne mill pollution are two key regulatory areas.

In the US, the lead executive agency for trade is the Office of the United States Trade Representative <www.ustr.gov>, a cabinet-level position that shares policy jurisdiction with the departments of Commerce and Treasury. When it comes to unfair trade remedies for steel imports, a key agency is the International Trade Administration <www.ita.doc.gov>, a branch of the Department of Commerce which plays a central role in screening steel import complaint petitions. It shares responsibility with the quasi-judicial ITC <www.usitc.gov>, which determines injury impacts from dumping and subsidies, proposes remedies, and conducts safeguard investigations.

The dominant trade union is the USWA. Its Canadian division, based in Toronto, is semi-autonomous <www.uswa.ca> within the broader continental union <www.uswa.org>.

Among trade periodicals, *New Steel* provided lively coverage of the mini-mill segment until the journal's demise in September 2001. Its site <www.newsteel.com> continues to offer full article and news coverage from 1996-2001. The *CRU Monitor — Steel* provides very useful background on the world market conditions <www.crumonitor.com>. The on-line newsletter <www.steelnews.com> records major industry events in North America and globally. The public section of the SteelVillage website <www.steelvillage.com> is also helpful in following industry events. A number of US policy think tanks have devoted attention to the steel question. These include the Institute of International Economics <www.iie.com> and the Cato Institute's Centre for Trade Policy Studies <www.cato.org>.

Airline Transport

With over 70 per cent of the scheduled passenger traffic, Air Canada <www.aircanada.ca> has for several years been the dominant corporate presence. The recent proliferation of niche carriers within this corporate family, including Tango, Jazz, Jetz, and Zip, can all be accessed from the main Air Canada group site. Other commercial airlines include Westjet <www.westjet.com>, Canjet <www.canjet.com>, Jetsgo <www.jetsgo.net>, and Air Transat <www.airtransat.com>. Since its privatization in the late 1990s, the air traffic control system NavCanada <www.navcanada.ca> has been part of the commercial setting. The same can be said, increasingly, about major airport facilities across the country, many of which are represented by the Canadian Airports Council <www:cacairports.ca>.

The Air Transport Association of Canada <www.atac.ca> is the main trade association for carriers, representing dozens of smaller players across the air transport sector. Its opposite number in the US is the Air Transport Association of America <www.airlines.org>. For the global industry, the International Air Transport Association <www.iata.org> is a leading voice. The retail ticketing sector established the Association of Canadian Travel Agents <www.acta.net> in 1977. The Canadian Business Aviation Association <www.cbaa.ca> represents the operators of business aircraft (some 50 members).

As the CAI dramas revealed, there are a large number of trade unions involved in the air transport sector. These include the Canadian Auto Workers <www.caw.ca>, the International Association of Machinists and Aerospace Workers <www.iamaw.org>, the Airline Pilots Association <www.alpa.org>, and the Union of Canadian Transportation Employees within the Public Service Alliance <www.psac.ca>.

Air transport is a federal policy jurisdiction. The Transport Canada site <www.tc.gc.ca> offers information on departmental operations, together with policy background on airport authorities. The regulatory agency is the CTA <www.cta-otc.gc.ca>. As noted above, the Competition Bureau <www.ic.gc.ca> holds responsibilities for merger review and competitive practice enforcement and has emerged as a key player in commercial air policy. In 2000-02 the reports by Debra Ward, the Independent Transition Observer on Airline Restructuring <www.tc.gc.ca>, offered useful perspectives on post-merger events. On safety matters, the Transport Safety Board of Canada <www.tsb.gc.ca> is the national investigator. Both the Aeronautics Act and Regulations are available online <www.tc.gc.ca/Actsregs/aviation_e.ht>. When it comes to international air service agreements, the Department of Foreign Affairs and International Trade <www.dfait-maeci.gc.ca> shares the file with Transport.

In the US, the Department of Transport <www.dot.gov> is the bureau of record, while the Federal Aviation Administration <www1.faa.gov> oversees safety standards.

As an agency of the United Nations, the International Civil Aviation Organization <www.icao.int> is an important foundation of the international air transport regime.

Several Canadian air-sector trade publications offer partial access online. They include *Wings* <www.wingsmagazine.com> and *Canadian Aviation* <www.canadianaviation.com>. For the global industry, *Flight International* <www.flightinternational.com> provides valuable coverage. The *Journal of Air Transport Management*, available through ScienceDirect in many academic libraries, provides very timely perspectives on the contemporary airline business.

References

Adams, Walter, and Hans Mueller. 1982. "The Steel Industry." In Walter Adams, ed., *The Structure of American Industry*. 6th ed. New York, NY: Macmillan. 73-135.

Ahlbrandt, Roger S., Richard J. Fruehan, and Frank Giarratani. 1996. *The Renaissance of American Steel*. New York, NY: Oxford University Press.

Alfano, James. 1997 (3 December). "Economic Forecast '98." Remarks to Hamilton and District Chamber of Commerce.

——. 2002 (1 May). "Address at Stelco Inc. 92nd Annual Shareholders' Meeting."

Algoma Steel Inc. 1986. *Annual Report*. Sault Ste. Marie, ON: Algoma Steel Inc.

Allison, Graham. 1971. *The Essence of Decision: Explaining the Cuban Missile Crisis*. Boston, MA: Little Brown..

APEC. 1974. *Steelmaking in the Atlantic Provinces*. Halifax, NS: Atlantic Provinces Economic Council.

Arnold, Gerry. 1986. "Premier to push for minimum 80,000 for rail order for Sysco." *The Chronicle-Herald* (20 November): 1, 28.

——. 1987. "PM, premier talk Sysco today." *The Chronicle-Herald* (16 January): 1,16.

Ashley, C.A. 1963. *The First Twenty-Five Years: A Study of Trans-Canada Airlines*. Toronto, ON: Macmillan.

Atkinson, Michael M., and William D. Coleman. 1989. *The State, Business and Industrial Change in Canada*. Toronto, ON: University of Toronto Press.

Avmark. 1993. "The Silent Conversations Issue: Farce or Tragedy?" *Avmark Aviation Journal* (May).

Axworthy, Lloyd. 1984 (10 May). *A New Canadian Air Policy*. Ottawa, ON.

Bakvis, Herman. 1991. *Regional Ministers*. Toronto, ON: University of Toronto Press.

Baldwin, John. 1975. *The Regulatory Agency and the Public Corporation*. Cambridge, MA: Ballinger.

Barnett, D.F., and L. Schorsch. 1983. *Steel: Upheaval in a Basic Industry*. Cambridge, MA: Ballinger.

Baron, David P. 1999. "Integrated Market and Non-Market Strategies in Client and Interest Group Politics." *Business and Politics* 1(1): 7-34.

Bishop, Joan. 1990. "Sydney Steel: Public Ownership and the Welfare State, 1967 to 1975." In Kenneth Donovan, ed., *The Island: New Perspectives on Cape Breton's History, 1713-1990*. Fredericton, NB: Acadiensis Press. 165-86.

Borrus, Michael. 1983. "The Politics of Competitive Erosion in the US Steel Industry." In John Zysman and Laura Tyson, eds., *American Industry and International Competition*. Ithaca, NY: Cornell University Press. 60-105.

Bothwell, Robert, and William Kilbourn. 1979. *C.D. Howe: A Biography*. Toronto, ON: McClelland and Stewart.

Brieger, Peter. 2002. "Just in time deliveries, the Japanese way." *National Post* (29 June).

——. 2003. "Stelco to beg Ottawa for steel tariff." *National Post*. (20 March): FP6.

Brown, Anthony E. 1987. *The Politics of Airline Deregulation*. Knoxville, TN: University of Tennessee Press.

Bruck, Connie. 1989. *The Predator's Ball: The Junk Bond Raiders and the Man Who Staked Them*. New York, NY: Simon and Schuster.

Button, Kenneth. 1989. "Liberalizing the Canadian Scheduled Aviation Market." *Fiscal Studies*: 19-52.

Campbell, Clayton. 1987. "Sysco-CN deal sealed." *The Chronicle-Herald* (14 January): 1,2.

Campbell, Robert, and Leslie A. Pal. 1994. "Air Farce: Airlines Policy in a Deregulatory Environment." In *The Real Worlds of Canadian Politics*. 3rd ed. Peterborough, ON: Broadview Press. 83-141.

Canada. 1957. Tariff Board. *Primary Iron and Steel*. Ref. 118. Ottawa, ON: Queen's Printer.

——. 1974. Hon. Willard J. Estey. *Steel Profits Inquiry*. Ottawa, ON: Information Canada.

——. 1975. Ministry of Transport. *Air Canada Inquiry Report*. Ottawa, ON: Information Canada.

——. 1978a. Industry, Trade and Commerce. *Report of the Sector Task Force on Canada's Forest Products Industry*. Ottawa, ON.

——. 1978b. Industry, Trade and Commerce. *Review of the Canadian Forest Products Industry*. Ottawa, ON.

——. 1978c (December). Industry, Trade and Commerce. *Report of the Sector Task Force on the Canadian Primary Iron and Steel Industry*. Ottawa, ON.

——. 1979. Industry, Trade and Commerce. *Response of the Federal Government to the Recommendations of the Consultative Task Force on the Canadian Forest Products Industry*. Ottawa, ON.

——. 1981. Economic Council of Canada. *Reforming Regulation*. Ottawa, ON.

——. 1982. House of Commons. Standing Committee on Transport. *Domestic Air Carrier Policy*. Ottawa, ON.

——. 1985. Ministry of Transport. *Freedom to Move*. Ottawa, ON.

——. 1986. Law Reform Commission of Canada. *Policy Implementation, Compliance and Administrative Law*. Ottawa, ON: Supply and Services Canada.

——. 1991. House of Commons. Canada-US Air Transport Agreement Special Committee. *Report*. Ottawa, ON.

——. 1993. Auditor General of Canada. *Annual Report, 1993*. Ottawa, ON.

——. 1996a. Industry Canada. *Primary Steel: Overview and Prospects*. <strategis.ic.gc.ca/SSG/mm01070e.htm>.

——. 1996b (March). Revenue Canada. Anti-Dumping and Countervail Directorate. Ottawa, ON.

——. 1997a. Industry Canada. *Forest Products: Overview and Prospects*. <strategis.ic.gc.ca/SSG/fb01002e.htm>.

——. 1997b. Industry Canada. *Canadian Market Pulp Industry*.

<strategis.ic.gc.ca/SSG/fbo1043e.htm>.

——. 1999a. Commissioner of Competition. Industry Canada. *Competition Issues Related to Air Restructuring*. Ottawa. September. <strategis: ic.gc.ca/SSG/cto1638e.htm>.

——. 1999b. Transport Canada. *A Policy Framework for Domestic Airline Restructuring in Canada*. Ottawa. October. <www.tc.gc.ca/pol/en>.

——. 1999c (December). House of Commons. Standing Committee on Transport. *Restructuring Canada's Airline Industry: Fostering Competition and Protecting the Public Interest*. Ottawa, ON.

——. 2002a. Industry Canada. *Forest Products: Overview and Prospects*. <strategis.ic.gc/SSG/fbo1005e.htm>.

——. 2002b (August). Canadian International Trade Tribunal. *Safeguard Inquiry into the Importation of Certain Steel Goods*. No.GC-2001-001.

——. 2002c (September). Independent Transition Observer on Airline Restructuring. Debra Ward. *Airline Restructuring in Canada: Final Report*. Ottawa, ON.

——. 2002d. Industry Canada. Canadian Industrial Statistics. NAICS 322. <strategis.ic.gc/canadian_industry_statistics.cis.nsf/IDE/cis322defe.html>

——. Canadian International Trade Tribunal. *Annual Reports*. 1997/98 to 2001/02.

Canadian Forum. 1971. *A Citizen's Guide to the Gray Report*. Toronto, ON: New Press.

Carrere, Ricardo, and Larry Lohmann. 1996. *Pulping the South: Industrial Tree Plantations and the World Paper Economy*. London, UK: Zed Books.

Carson, Rachel. 1962. *Silent Spring*. Cambridge, MA: Riverside Press.

CBC. 1993 (19 August). *Radio Noon*.

Chandler, Alfred. 1962. *Strategy and Structure*. Cambridge, MA: MIT Press.

——. 1977. *The Visible Hand*. Cambridge, MA: Belknap Press.

Chronicle-Herald. 2002. "Stora Enso executives flown in." 17 May: C3.

Clancy, Peter. 1992. "The Politics of Pulpwood Marketing in Nova Scotia, 1960-85." In L. Anders Sandberg, ed., *Trouble in the Woods: Forest Policy and Social Conflict in Nova Scotia and New Brunswick*. Fredericton, NB: Acadiensis Press. 142-67.

——. 2001. "Atlantic Canada: The Politics of Public and Private Forestry." In Michael Howlett, ed., *Canadian Forest Policy: Adapting to Change*. Toronto, ON: University of Toronto Press. 205-36.

Clancy, Peter, and L. Anders Sandberg. 1995. "Crisis and Opportunity: The Political Economy of the Stora Mill Closure Controversy." *Business Strategy and the Environment* 4(4): 208-19.

Clancy, Peter, James Bickerton, Rodney Haddow, and Ian Stewart. 2000. *The Savage Years: The Perils of Reinventing Government in Nova Scotia*. Halifax, NS: Formac Publishing.

Clarkson, Stephen, and Christina McCall. 1994. *Trudeau and Our Times: The Heroic Delusion*. Vol.2. Toronto, ON: McClelland and Stewart.

Cockerill, Anthony. 1974. *The Steel Industry: International Comparison of Industrial Structure and Performance*. Cambridge, MA: Cambridge University Press.

Coleman, William D. 1988. *Business and Politics*. Montreal, QC: McGill-Queen's University Press.

Cook, Peter. 1981. *Massey at the Brink*. Toronto, ON: Collins.

Cooke, Henry. 2001. *Iron and Steel Works of the World*. 14th ed. London, UK: Metal Bulletin Books.

Corbett, D. 1965. *Politics and the Airlines*. Toronto, ON: University of Toronto Press.

Corman, June, Meg Luxton, D.W. Livingstone, and Wally Seccombe. 1993. *Recasting Steel Labour*. Halifax, NS: Fernwood Publishing.

CP News. 1987. "Full Sysco lunch pails leave Algoma in the slag." *The Chronicle-Herald* (23 January): 20.

——. 1989. "US slaps duties on steel rails." *The Chronicle-Herald* (25 August): 3.

CRU Monitor — Steel. 2002a (May). "The Need for Consolidation."

——. 2002b (October). "Chasing the Dragon."

CSPA. 1999. *CSPA Policy Agenda, 1999-2000*. Ottawa, ON: Canadian Steel Producers Association. <www.canadiansteel.ca>.

——. 2002. *Free Trade With Effective Trade Remedies*. Ottawa, ON: Canadian Steel Producers Association. <www.canadiansteel.ca>.

de Silva, K.E.A. 1988. *Pulp and Paper Modernization Grants Program — An Assessment*. Ottawa, ON: Economic Council of Canada.

Dempsey, Paul Stephen, and Andrew R. Goetz. 1992. *Airline Deregulation and Laissez-Faire Mythology*. Westport, CT: Quorum Books.

Dobuzinskis, L. 1996. "Trends and Fashions in the Marketplace of Ideas." In L. Dobuzinskis, M. Howlett, and D. Laycock, eds., *Policy Studies in Canada: The State of the Art*. Toronto, ON: University of Toronto Press. 91-124.

Doern, G. Bruce. 1995. "Sectoral Green Politics: Environmental Regulation and the Canadian Pulp and Paper Industry." *Environmental Politics* 4(2): 219-43.

Doern, G. Bruce, and Thomas Conway. 1994. *The Greening of Canada: Federal Institutions and Decisions*. Toronto, ON: University of Toronto Press.

Doern, G. Bruce, and Brian Tomlin. 1991. *Faith and Fear*. Toronto, ON: Stoddart.

Doganis, Rigas. 1991. *Flying Off Course: The Economics of International Airlines*. 2nd ed. London, UK: Harper Collins Academic.

——. 2001. *The Airline Business in the 21st Century*. London, UK: Routledge.

Domtar Inc. 2001. *Annual Report*. Montreal, QC: <www.domtar.com>.

Donald, W.J.A. 1915. *The Canadian Iron and Steel Industry*. Boston, MA: Houghton Mifflin.

Dufresne, Yves. 2003. "The Canada/ US Air Service Relationship: Time for Further Liberalization." Toronto, ON: Air Policy Forum. <www.cacairports.ca>.

Eden, Lorraine. 1991. "Bringing the Firm Back In: Multinationals in International Political Economy." *Millennium: A Journal of International Studies* 20(2): 197-224.

Eichner, Alfred S. 1985. *Toward a New Economics*. New York, NY: Macmillan.

Elliot, Geoffrey. 2003. "Adapting to Change: Canada's International Airline Policy." Toronto, ON: Air Policy Forum. <www.cacairports.ca>.

Erwin, Steve. 2002. "Mixed ruling delivered on steel imports." *The Chronicle-Herald* (6 July): D28.

Feldman, Joan M. 1996 (5 May). "Some Call it Oligopoly." *Air Transport World*.

Ferguson, Thomas. 1984. "From Normalcy to the New Deal: Industrial Structure, Party Competition, and American Public Policy in the Great

Depression." *International Organization* 38(1): 41-94.

Ferguson, Thomas, and Joel Rogers, eds., 1981. *The Hidden Election: Politics and Economics in the 1980 Presidential Campaign.* New York, NY: Pantheon Books.

Ferguson, Thomas, and Joel Rogers. 1986. *Right Turn: The Decline of the Democrats and the Future of Republican Politics.* New York, NY: Hill and Wang.

Financial Post Business. 2003 (June). FP 500.

Finkel, Alvin. 1979. *Business and Social Reform in the Thirties.* Toronto, ON: James Lorimer.

Fitzpatrick, Peter, and Ian Jack. 1999. "Ottawa paves way for airline restructuring." *National Post* (14 August): 1, 2.

FPAC. 2001. *Reducing Effluent.* Ottawa, ON: Forest Products Association of Canada. <www.opendoors.cppa.ca>.

FSAC. 1986 (December). *Planning for Change.* Ottawa, ON: Forest Sector Advisory Council.

Forster, Ben. 1986. *A Conjuncture of Interests.* Toronto, ON: University of Toronto Press.

Frank, David. 1977. "The Cape Breton Coal Industry and the Rise and Fall of the British Empire Steel Corporation." *Acadiensis* 7(1): 3-34.

Freeman, Bill. 1982. *1005: Political Life in a Union Local.* Toronto, ON: James Lorimer.

French, Richard. 1980. *How Ottawa Decides: Planning and Industrial Policy-Making 1968-80.* Toronto, ON: James Lorimer.

Galbraith, J.K. 1978. *The New Industrial State.* 3rd ed. rev. New York, NY: Signet.

Gialloreto, Louis. 1988. *Strategic Airline Management: "The Global War Begins."* London, UK: Pitman.

——. 1994. "No Space for the Middle of the Roader." *Avmark Aviation Economist* (September).

Gibbens, Robert. 1999. "Quebec court ruling ends takeover bid." *National Post* (6 November): D1, 5.

Gillen, David W., Tae H. Oum, and Michael W. Trethaway. 1988a. "Entry Barriers and Anti-Competitive Behaviour in a Deregulated Airline Market: The Case of Canada." *International Journal of Transport Economics* 15(1): 29-41.

Gillen, David W., W.T. Stanbury, and Michael W. Trethaway. 1988b. "Duopoly in Canada's Airline Industry: Consequences and Policy Issues." *Canadian Public Policy* 14(1): 15-31.

Gillis, R.P., and T.R. Roach. 1986. *Lost Initiatives.* Westport, CT: Greenwood Press.

Globe and Mail. 2001. "Corporate brass faces the new year with worries." 2 January: B1.

Goldenberg, Susan. 1984. *Canadian Pacific: A Portrait of Power.* Toronto, ON: Methuen.

——. 1994. *Troubled Skies: Crisis, Competition and Control in Canada's Airline Industry.* Whitby, ON: McGraw-Hill Ryerson.

Gollner, Andrew. 1983. *Social Change and Corporate Strategy.* Stamford, CT: Issue Action Publications.

Gould, Douglas. 2003. "Steely resolve needed." *Globe and Mail* (4 August): B8.

Grant, Wyn. 1990. "Forestry and Forest Products." In William D. Coleman and Grace Skogstad, eds., *Policy Communities and Public Policy in Canada.* Toronto, ON: Copp Clark Pitman. 118-40.

Green, Christopher. 1990. *Canadian Industrial Organization and Policy.* Scarborough, ON: McGraw-Hill.

Greenpeace International. 1990. *The Greenpeace Guide to Paper*. Vancouver, BC.

——. 1998. "Chlorine free papermaking." <www.greenpeace.org/~toxics>.

Greenspon, Lawrence. 1988. "After 700 years, Stora throws a party." *Globe and Mail* (17 June): B1, 2.

Guthrie, John A. 1941. *The Newsprint Paper Industry*. Cambridge, MA: Harvard University Press.

Hallvarsson, Mats. 1987. "The Jewel of the Kingdom." *Sweden Works*. Halmsted, SWE: Civilien AB. 80-90.

Hanlon, Pat. 1996. *Global Airlines: Competition in a Transnational Industry*. Oxford: Butterworth-Heinemann.

Harris, Ralph F. 1978. "The Regulation of Air Transportation." In D. Bruce Doern (ed.) *The Regulatory Process in Canada*. Toronto, ON: Macmillan of Canada. 212-36.

Harrison, Bennett. 1994. *Lean and Mean: The Changing Landscape of Corporate Power in the Age of Flexibility*. New York, NY: Basic Books.

Harrison, Kathryn. 1996a. *Passing the Buck: Federalism and Canadian Environmental Policy*. Vancouver, BC: UBC Press.

——. 1996b. "The Regulator's Dilemma: Regulation of Pulp Mill Effluents in the Canadian Federal State." *Canadian Journal of Political Science* 29(3): 469-96.

——. 1999. "Retreat From Regulation: The Evolution of the Canadian Environmental Regulatory Regime." In G.B. Doern, ed., *Changing the Rules: Canadian Regulatory Regimes and Regulations*. Toronto, ON: University of Toronto Press. 122-42.

——. 2002. "Ideas and Environmental Standard-Setting: A Comparative Study of Regulation of the Pulp and Paper Industry." *Governance* 15(1): 65-96.

Hart, Michael. 2002. *A Trading Nation*. Vancouver, BC: UBC Press.

Harvey, David. 1989. *The Conditions of Postmodernity: An Enquiry into the Origins and Cultural Change*. Oxford: Basil Blackwell.

Hawes, Michael K. 1986. "The Steel Industry: Change and Challenge." *International Journal* 42: 25-58.

Henderson, Jennifer. 1992. "Sydney Steel up for grabs." *Financial Post* (25-27 January): 4.

Henwood, Doug. 1998. *Wall Street*. London: Verso.

Heron, Craig. 1988. *Working in Steel*. Toronto, ON: McClelland and Stewart.

Hirschmann, Albert O. 1958. *The Strategy of Economic Development*. New Haven, CT: Yale University Press.

Hoare, Eva. 1986. "CP urged to aid Sysco." *The Chronicle-Herald* (24 November): 1, 20.

Hoerr, John P. 1988. *And the Wolf Finally Came*. Pittsburgh, PA: University of Pittsburgh Press.

Hogan, William T. 1983. *World Steel in the 1980s: A Case of Survival*. Lexington, KY: Heath.

Howlett, Michael, and M. Ramesh. 1995. *Studying Public Policy*. Toronto, ON: Oxford University Press.

Hufbauer, Gary, and Ben Goodrich. 2001. "Steel: Big Problems, Better Solutions." Institute for Interational Economics. *International Economic Policy Briefs* 1-9 (July).

Humphreys, Barry. 1994. "Do Airlines Still Need to Own CRSs?" *Avmark Aviation Journal* (April).

Hutton, Will. 1996. *The State We're In*. London: Vintage.

International Iron and Steel Institute. 2003. *World Steel in Figures. 2003*

Edition. Brussels. <www.worldsteel.org>.

Inwood, Kris. 1987. "The Iron and Steel Industry." In Ian M. Drummond, ed., *Progress Without Planning*. Toronto, ON: University of Toronto Press. 185-207.

IPSCO Inc. 2001. *Annual Report*. Regina: <www.ipsco.com>.

Janisch, H.N. 1979. *The Regulatory Process of the Canadian Transport Commission*. Ottawa, ON: Law Reform Commission of Canada.

Jeffers, Alan. 1986. "Young backtracks on Grits." *The Chronicle-Herald* (7 November): 1,28.

———. 1990. "New improved Sysco considered saleable." *The Chronicle-Herald* (12 May): B3.

Jones, Kent O. 1986. *Politics v. Economics in the World Steel Trade*. London: Allen and Unwin.

Jordan, W.A. 1983. *The Performance of Regulated Canadian Airlines in Domestic and Transborder Operations*. Ottawa, ON: Consumer and Corporate Affairs.

Junkerman, John. 1987. "Blue Sky Management: The Kawasaki Story." In Richard Peet, ed., *International Capitalism and Capitalist Restructuring*. Boston, MA: Allen and Unwin. 131-44.

Kahn, Alfred E. 1988. "Surprises of Airline Deregulation" *AEA Papers and Proceedings* 7/8(2): 316-22.

Kearns, David T., and Mark Nagler. 1992. *Prophets in the Dark*. New York, NY: Harper Business.

Keenan, Greg. 2002 "Steelmakers predict recovery." *Globe and Mail* (2 February): B2.

Keith, Ronald A. 1972. *Bush Pilot With a Briefcase*. Toronto, ON: Doubleday Canada.

Kenny, Jim. 1997. "I Bet You Next Year's Profits the Price of Pulp Goes Up." *Pulp and Paper International* (September).

Kilbourn, William. 1961. *The Elements Combined: A History of the Steel Company of Canada*. Toronto, ON: Clarke Urwin.

Kingdon, John W. 1984. *Agendas, Alternatives and Public Policy*. Boston, MA: Little Brown.

Kurth, James R. 1979. "The Political Consequences of the Product Cycle." *International Organization* 33(1):1-34.

Kymlicka, B.B. 1987. "Canadian Steel in Washington, 1984." Monograph. London, ON: University of Western Ontario.

Langford, John W. 1981. "Air Canada." In Allan Tupper and G. Bruce Doern, eds., *Public Corporations and Public Policy in Canada*. Montreal, QC: Institute for Research on Public Policy. 251-84.

Langford, John, and Ken Huffman. 1988. "Air Canada." In Allan Tupper and G. Bruce Doern, eds., *Privatization, Public Policy and Public Corporations in Canada*. Montreal, QC: Institute for Research on Public Policy. 93-150.

Langille, David. 1987. "The BCNI and the Canadian State." *Studies in Political Economy* (Autumn): 41-85.

Lazar, Fred. 1984. *Deregulation of the Canadian Airline Industry*. Toronto, ON: Key Porter.

Levitt, Theodore. 1965. "Exploit the Product Life Cycle." *Harvard Business Review* 45: 81-94.

Lindeberg, Anders. 1996. "A Little Insurance for the Future of Pulp." *Pulp and Paper International* (September).

Lissitzyn, O.J. 1964. "Bilateral Agreements in Air Transport." *Journal of Air Law and Commerce* 30: 248-63.

Litvak, Isaiah A. 1982. "National Trade Associations: Business-Government Intermediaries." *Business Quarterly* 47(3): 31-42.

Litvak, Isaiah A., and Christopher J. Maule. 1977. *Corporate Dualism and the Canadian Steel Industry*. Ottawa, ON: Supply and Services Canada.

———. 1985. "The Canadian Aluminum and Steel Industries." In D.G. McFetridge, ed., *Technological Change in Canadian Industry*. Toronto, ON: University of Toronto Press. 145-75.

Luczak, Helen. 1985. "The Steel Tariff of 1957." Unpublished paper. London, ON: University of Western Ontario.

MacArthur, Keith, and Andrew Willis. 2003. "Air Canada creditors up next." *Globe and Mail* (2 June): B1.

MacDonald, Alena. 2003. "Stora expansion gets approval." *The Reporter* (31 January): 1,2.

MacDonald, Dan. 1986a. "Fewer orders for Sysco." *The Chronicle-Herald* (19 November): 1, 28.

———. 1986b. "Sysco prospects dim." *The Chronicle-Herald* (26 November): 1, 24.

———. 1987. "Cabinet group to study Sysco plan." *The Chronicle-Herald* (19 March): 1, 32.

MacDonald, Dan, and Gerry Arnold. 1987. "Sysco's ball in N.S. court." *The Chronicle-Herald* (14 April): 1, 16.

Macdonald, Doug. 1991. *The Politics of Pollution*. Toronto, ON: McClelland and Stewart.

MacDowell, Duncan. 1984. *Steel at the Sault*. Toronto, ON: University of Toronto Press.

Madrick, Jeff. 1996 (April). "How to Succeed in Business." *New York Review of Books*. 43(7).

Magaziner, Ira, and Robert Reich. 1982. *Minding America's Business*. New York, NY: Harcourt, Brace and Jovanovich.

Mahon, Rianne. 1977. "Canadian Public Policy: The Unequal Structure of Representation." In Leo Panitch, ed., *The Canadian State: Political Economy and Political Power*. Toronto, ON: University of Toronto Press. 165-98.

———. 1979. "Captive Agents or Hegemonic Apparatuses." *Studies in Political Economy* 1: 162-200.

Main, J.R.K. 1967. *Voyageurs of the Air*. Ottawa, ON: Queen's Printer.

March, James G. 1962. "The Business Firm as a Political Coalition." *Journal of Politics* 24(2): 662-78.

Marchak, Patricia. 1982. *Green Gold: The Forest Industry in British Columbia*. Vancouver, BC: University of British Columbia Press.

———. 1995. *Logging the Globe*. Montreal, QC: McGill-Queen's University Press.

Markusen, Ann Roell. 1985. *Profit Cycles, Oligopoly and Regional Development*. Cambridge, MA: MIT Press.

Masi, Anthony C. 1991. "Structural Adjustment and Technological Change in the Canadian Steel Industry, 1970-1986." In Daniel Drache and M.S. Gertler, eds., *The New Era of Global Competition*. Montreal, QC: McGill-Queen's University Press. 181-205.

Mathias, Phillip. 1971. *Forced Growth*. Toronto, ON: James Lewis and Samuel.

May, Elizabeth. 1982. *Budworm Battles*. Halifax, NS: Four East Publications.

———. 1998. *At the Cutting Edge*. Toronto, ON: Key Porter.

McBrearty, Lawrence. 2002. "Tribunal Handling of Steel Crisis all Wrong." <www.uswa.ca/eng/news_>

McBride, Stephen. 2001. *Paradigm Shift: Globalization and the Canadian State*. Halifax, NS: Fernwood Publishing.

McCann, L.D. 1981 "The Mercantile-Industrial Transition in the Metal Towns of Pictou County, 1857-1931." *Acadiensis* 10(2): 29-64.

McConnell, Grant. 1963. *Steel and the Presidency, 1962*. New York, NY: W.W. Norton.

McGrath, T.M. 1992. *History of Canadian Airports*. Ottawa, ON: Lugus Publications.

Melnbardis, Robert. 1992. "Domtar moves back from the abyss." *Financial Times* (12 December): 7.

Metal Bulletin. 2000. *Mini Mill Data Book*. 2nd ed. Surrey, UK: Metal Bulletin Books.

Micklethwaite, John, and Adrian Wooldridge. 1996. *The Witch Doctors*. New York, NY: Times Business.

Miliband, Ralph. 1973. *The State in Capitalist Society*. London: Quartet Books.

Morris, Dick. 1997. *Behind the Oval Office*. New York, NY: Random House.

Morton, Peter. 2003. "WTO rules against US steel barriers." *National Post* (11 November): FP2.

Myrden, Judy. 2002. "Stora Enso ups the ante." *The Chronicle-Herald* (15 March): C1.

Naylor, R.T. 1975. *The History of Canadian Business, 1867-1914*. 2 vols. Toronto, ON: James Lorimer.

——.1987. *Hot Money*. Toronto, ON: McClelland and Stewart.

Nelles, H.V. 1974. *The Politics of Development*. Toronto, ON: Macmillan.

Newman, Peter C. 1975. *The Canadian Establishment*. Toronto, ON: McClelland and Stewart.

Niosi, Jorge. 1985. *Canadian Multinationals*. Toronto, ON: Garamond Press.

Nova Scotia. 1968. Voluntary Economic Planning. *Sydney Steelmaking Study*. Halifax, NS.

——. 1975. Auditor-General. *Report*. Halifax, NS.

NSFI. 1972. *The New Forest*. Grycksbo, SWE: Nova Scotia Forest Industries Ltd.

——. 1983. *Submission to the Royal Commission on Forestry*. Port Hawksbury, NS: Nova Scotia Forest Industries Ltd.

Olson, Mancur. 1971. *The Logic of Collective Action*. Cambridge, MA: Harvard University Press.

Ontario. 1988. Premier's Council. *Competing in the New Global Economy*. Toronto, ON: Queen's Printer for Ontario.

——. 1978. *Report of the Special Task Force on the Ontario Pulp and Paper Industry*. Toronto, ON: Ministry of Industry and Tourism.

Osbaldeston, Gordon. 1988. "Dear Minister." *Policy Options Politiques* 9: 3-11.

Oum, T.H., W.T. Stanbury, and Michael W. Trethaway. 1991. "Airline Deregulation in Canada and Its Economic Effects." *Transportation Journal* 30(4): 4-22.

Pal, Leslie A. 1992. *Public Policy Analysis*. Scarborough, ON: Nelson Canada.

Parenteau, William, and L. Anders Sandberg. 1995. "Conservation and the Gospel of Economic Nationalism: The Canadian Pulpwood Question in Nova Scotia and New Brunswick: 1918-1925." *Environmental History Review*, 19(2): 57-83.

Parlour, James W. 1981. "The Politics of Water Pollution Control: A Case Study of the Canadian Fisheries Act Amendments and the Pulp and Paper Effluent Regulations, 1970." *Journal of Environmental Management* 13: 127-49.

Perl, Anthony. 1994. "Public Enterprise as an Expression of Sovereignty." *Canadian Journal of Political Science* 27(1): 23-52.

Phillips, Roger. 2001 (August). "The Value of 201." *New Steel*. 17(8).

Pigott, Peter. 1998. *Wingwalkers: A History of Canadian Airlines International*. Vancouver, BC: Harbour Publishing.

Pooley, Frank. 1996. *Air Canada, A Pocket Guide: The History of Air*

Canada From 1937. Burnham, UK: FJP Photo Services.

Potter, Kent. 1986. "Carriers seek new routes to help bolster bottom line." *The Financial Post* (29 March): 48.

Poulantzas, Nicos. 1975. *Political Power and Social Classes.* London, UK: Verso.

Pratt, Larry, and Ian Urquhart. 1994. *The Last Great Forest.* Edmonton, AB: NeWest Press.

Pressman, Jeffrey B., and Aaron B. Wildavsky. 1973. *Implementation.* Berkeley, CA: University of California Press.

Price, David. 1991. "Company Profile: Europe's Largest Paper Group, Stora Feldmuhle, Consolidates." *EIU Paper and Packaging Analyst* (November): 48-59.

Raddall, Thomas H. 1979. *The Mersey Story.* Liverpool, UK: Bowater Mersey Paper Co. Ltd.

Reich, Robert. 1990. "Who is Us?" *Harvard Business Review* 68(1): 53-64.

Regina v. Howard Smith Paper Mills Ltd., et al. 1954. *Ontario Reports* [1954]: 543-646.

———. 1955. *Ontario Reports* [1955]: 713-737.

Reschenthaler, G.B., and W.T. Stanbury. 1983. "Deregulating Canada's Airlines; Grounded by False Assumptions." *Canadian Public Policy* 9(2): 210-22.

Rhodes, Carolyn. 1993. *Reciprocity, US Trade Policy and the GATT Regime.* Ithaca, NY: Cornell University Press.

Romain, Ken. 1989. "US slaps Sysco with penalty duty." *Globe and Mail* (25 February): B1,4.

Sampson, Anthony. 1984. *Empires of the Sky.* London, UK: Hodder and Stoughton.

Sandberg, L. Anders, ed. 1992. *Trouble in the Woods: Forest Policy and Social Conflict in Nova Scotia and New Brunswick.* Fredericton, NB: Acadiensis Press.

Sandberg, L. Anders and John H. Bradbury. 1988. "Industrial Restructuring in the Canadian Steel Industry." *Antipode* 20(2): 102-121.

Sandberg, L. Anders, and Peter Clancy. 2000. *Against the Grain.* Vancouver, BC: University of British Columbia Press.

Savage, John, and Alan Bollard. 1990. *Turning it Around: Closure and Revitalization in New Zealand Industry.* Auckland, NZ: Oxford University Press.

Savage, Murray. 1977. *Domtar Ltd.* Ottawa, ON: Department of Supply and Services.

Sawatsky, John. 1987. *The Insiders: Government, Business and the Lobbyists.* Toronto, ON: McClelland and Stewart.

Schmidt, Lisa. 2002. "Feds to review steel imports." *The Chronicle-Herald* (23 March): B10.

Schumpeter, Joseph. 1939. *Business Cycles.* New York, NY: McGraw Hill.

Schwindt, Richard W. 1977. *The Existence and Exercise of Corporate Power.* Ottawa, ON: Supply and Services.

Scott, Ian. 1993. "Stora may close NS mills." *The Chronicle-Herald* (18 August): 1,2.

Shahery, Shaw. 2002. "The Global State of the Tissue Industry." *Pulp and Paper Canada* 103: 5.

Shenton, Harold. 1993. "Change, Challenge and Competition." *Avmark Aviation Journal* (September).

Shuman, James B., and David Rosenau. 1972. *The Kondratieff Wave.* New York, NY: Delta.

Sidor, Nicholas. 1981. *Forest Industry Development Policies: Industrial Strategy or Corporate Welfare.* Ottawa, ON: Canadian Centre for Policy Alternatives.

Sierra Legal Defence Fund. 2000. *Pulping the Law: How Pulp Mills Are Ruining Canadian Waters With Impunity*. Toronto, ON.

Sinclair, William F. 1990. *Controlling Pollution From Canadian Pulp and Paper Manufacturers: A Federal Perspective*. Ottawa, ON: Queen's Printer.

——. 1991. "Controlling Effluent Discharges from Canadian Pulp and Paper Manufacturers." *Canadian Public Policy* 17(1): 86-105.

Singer, Jacques. 1969. *Trade Liberalization and the Canadian Steel Industry*. Toronto, ON: University of Toronto Press.

Skene, Wayne. 1994. *Turbulence: How Deregulation Destroyed Canada's Airlines*. Vancouver, BC: Douglas and McIntyre.

Skeoch, L.A. 1966. *Restrictive Trade Practices in Canada*. Toronto, ON: McClelland and Stewart.

Smith, Phillip. 1986. *It Seems Like Only Yesterday*. Toronto, ON: McClelland and Stewart.

Solon, Daniel. 2002a (July). "Members Woes Threaten Top Teams." *Avmark Aviation Economist* 19(5).

——. 2002b (August). "Skyteam Maintains Momentum." *Avmark Aviation Economist*. 19(6).

Soros, George. 2000. *Open Society: Reforming Global Capitalism*. Boston, MA: Little Brown.

Soyez, Dietrich. 1988. "Stora Lured Abroad? A Nova Scotian Case Study in Industrial Decision-Making and Persistence." *The Operational Geographer* 16: 11-14.

Stanbury, W.T. 1992. *Business and Government Relations in Canada*. Scarborough, ON: Nelson Canada.

Stanbury, W.T., and F. Thompson. 1982. *Regulatory Reform in Canada*. Montreal, QC: Institute for Research on Public Policy.

Statistics Canada. 1997. *North American Industry Classification System*. Ottawa, ON.

Stevenson, Garth. 1987. *The Politics of Canada's Airlines*. Toronto, ON: University of Toronto Press.

Strange, Susan. 1986. *Casino Capitalism*. Oxford: Blackwell.

Stritch, Andrew J. 1991. "State Autonomy and Societal Pressure: The Steel Industry and US Import Policy." *Administration and Society* 23(3): 288-309.

Surrette, Ralph. 1975 (December). "Nova Scotia: Steel Industry Nears Collapse." *Last Post* 5(2).

Telmer, F.H. 1996a (May). "Our Big Neighbour." Canada Japan Business Committee Conference.

——. 1996b (October). "Progress and Obstacles in the Iron and Steel Trade." Latin American Iron and Steel Congress.

Tiffany, Paul A. 1988. *The Decline of American Steel: How Management, Labour and Government Went Wrong*. Oxford: Oxford University Press.

Traves, Tom. 1979. *The State and Enterprise: Canadian Manufacturers and the Federal Government, 1917-1931*. Toronto, ON: University of Toronto Press.

Trethaway, Michael. 2003. "The Broken Airline Business Model." Toronto, ON: Air Policy Forum. <www.cacairports.ca>

Troyer, Warner. 1977. *No Safe Place*. Toronto, ON: Clark Irwin.

Tupper, Allan. 1981. "Pacific Western Airlines." In Allan Tupper and G.B. Doern, eds., *Public Corporations and Public Policy in Canada*. Montreal, QC: Institute for Research on Public Policy. 285-318.

United States. 1992. Department of Transport. "Open Skies Statement."

USWA. 1996. United Steelworkers of America. "Trade Politics in the 1990s." Brief to the Foreign Affairs Sub-Committee and the Finance Sub-Committee on the Special Import Measures Act Review (25 November).

VanNijnatten, Debora L. 1998. "The Day the NGOs Walked Out." *Alternatives Journal* 24(2) (Spring): 10-15.

Verburg, Peter. 1998. "The Man of Steel is Fed Up" *Canadian Business* (26 June/ 10 July).

Vernon, Raymond. 1966. "International Investment and International Trade in the Product Cycle." *Quarterly Journal of Economics* 80(2): 190-207.

——. 1971. *Sovereignty at Bay: The Multinational Spread of US Enterprises.* New York, NY: Basic Books.

——. 1998. *In the Hurricane's Eye: The Troubled Prospects of Multinational Enterprises.* Cambridge, MA.: Harvard University Press.

Vogel, David. 1986. *National Styles of Regulation: Environmental Policy in Great Britain and the United States.* Ithaca, NY: Cornell University Press.

Waldie, Paul. 2003. "Cerberus's effort could succeed, experts say." *Globe and Mail* (29 November): B6.

Webb, Kernaghan. 1988. *Pollution Control in Canada: The Regulatory Approach of the 1980s.* Ottawa, ON: Law Reform Commission of Canada.

——. 1990. "Between Rocks and Hard Places: Bureaucrats, Law and Pollution Control." In Robert Paehlke and Douglas Torgerson, eds., *Managing Leviathan.* Peterborough, ON: Broadview Press. 201-27.

Weiss, Linda. 1998. *The Myth of the Powerless State.* Cambridge, UK: Polity Press.

Wells, Louis T. 1972. "Introduction." In Louis T. Wells (ed.), *The Product Life Cycle and International Trade.* Boston, MA: Harvard School of Business Administration.

Williams, George. 1995. "Frequent Flyer Programs." *Avmark Aviation Economist*: 4-17.

Woodbridge Reed and Associates. 1988. *Canada's Forest Industry: the Next Twenty Years.* 5 vols. Ottawa, ON.

World Steel Outlook. 1998. "IPSCO Inc." Quarter 4. Sheffield, UK: MEPS (Europe) Ltd.

For publicly listed companies, a comprehensive source of financial and historical information can be found on the CD-ROM by Compact Canada, *CanCorp Plus, Corporate Information on Canadian Companies.* Other valuable indexes to newspaper and magazine coverage are the *Canadian Business Periodicals Index* and the *Canadian News Index* (now *Canadian Index*).

Index